Date: 7/8/20

155.7 BAR
Barber, Nigel,
Evolution in the here and
now : how adaptation and

EVOLUTION IN THE HERE AND NOW

How Adaptation and Social Learning Explain Humanity

Nigel Barber

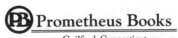

Guilford, Connecticut

Prometheus Books

An imprint of The Rowman & Littlefield Publishing Group, Inc.
4501 Forbes Boulevard, Suite 200
Lanham, Maryland 20706
www.rowman.com

Distributed by NATIONAL BOOK NETWORK

British Library Cataloguing in Publication Information Available

Library of Congress Cataloging-in-Publication Data Available

ISBN 9781633886186 (cloth)
ISBN 9781633886193 (ebook)

♾™ The paper used in this publication meets the minimum requirements of American National Standard for Information Sciences Permanence of Paper for Printed Library Materials, ANSI/NISO Z39.48-1992.

To Trudy,

my wife, muse, soulmate, best friend, and champion

CONTENTS

INTRODUCTION

This book introduces a new way of understanding the evolution of complex behavior for humans and other species. A skeptic might say that there is nothing new under the sun. There is always a grain of truth in that observation. Let me explain in broad strokes what is new here.

This book departs from the two main approaches to human behavior. The oldest is cultural determinism: maintaining that human societies differ due to socially transmitted ideas. This means that every society is arbitrarily different from others. Moreover, no human group is efficiently adapted to the way that they make their living or other features of the natural ecology. Being subject to internally driven cultural changes, human behavior is not predictable from the natural ecology in any but the most superficial ways, such as people who live in the Arctic making homes from snow and clothing from animal skins because these materials are available whereas wood and cotton are not.

Evolutionary psychology emerged as a revolt against the perceived arbitrariness of human behavior. If other species behave in ways that enhance their prospects of survival and reproduction, surely humans— products of the same process of evolution by natural selection—would do the same. Evolutionary psychologists emphasized adaptive patterns in behavior across most societies, such as men being more interested in physical attractiveness in a prospective spouse than women are, and women being more interested in the social status of a partner. They argued that such cross-culturally invariant patterns are evidence of Darwinian adaptation. Men are more interested in low-cost reproduction via "casual" sex whereas women need to ensure masculine investment in their offspring to have the best chance of raising them to maturity.

Although early evolutionary psychologists presented a refreshing challenge to the circularity and muddle of cultural determinism, their key claims were overstated. Gender differences that they assumed to be universal turned out to change over time. Gender differences in sexual psychology and behavior are disappearing in the modern world. This phenomenon pokes a hole in the argument that gender differences and other such cross-societal patterns are part of our genetic heritage and a product of gene-based adaptation.

Behavioral flexibility is not peculiar to humans. Many people fear snakes that could give fatal bites—at least for a few poisonous species. Fear of snakes is certainly powerful, but attributing it to our evolutionary history as hunter-gatherers may be a stretch. Indeed, fear of snakes in nonhuman primates is contingent on early experiences. If so, what humans fear may not be a consequence of atavistic experiences in hunter-gatherer societies. This impression is reinforced by remarkable research on the reaction of moose in North America to the reintroduction of wolves to their habitat after generations of extinction.

Initially, the moose showed no fear of their ancestral predators in a rather stunning disconfirmation of the idea that adaptive behavior is routinely transmitted via genes. Instead, moose acquired fear of wolves only after they were attacked. Young moose learned to fear the sight and scent of wolves by observing the impact of such stimuli upon their mothers. The notion that complex behavior, such as predator avoidance, is encoded in genes took a beating. Evolution is not restricted to genetic adaptation but works via learning and other mechanisms. In the end, natural selection has no preference for genes over learning. All that matters is whether the individual is successful in the current environment, not how their behavioral adaptation arose. The adaptiveness of moose reactions to their ancestral archpredators is really the consequence of their adjustment to the current environment. If so, then behavioral adaptation can occur in very short time scales rather than the millions of years it can take for anatomical adaptations to occur.

This impression is strongly confirmed by the novel antics of many species that share our habitat and have left the wild outdoors for city streets. In Thailand, long-tailed macaques happily use power lines as an elevated path over streets, as if they were jungle vines. They are skilled at stealing food from market stalls and even break into houses and steal from refrigerators. This means that they develop the capacity to eat scores of novel foods. In one of their weirdest adaptations to city life, they floss their teeth and use stolen strands of human hair as dental floss. These amusing novel behaviors were recounted by George McGavin in the British Broadcasting Corporation documentary series *Monkey Planet*. That these behaviors can emerge in a single generation in response to current opportunities shows how very far off the mark evolutionary psychologists were in assuming that adaptive behavior is genetically heritable and fixed.

The assumption that complex human behavioral propensities are encoded in genes is equally disputable. This is particularly true of gender differences in psychology that could only reside in the sex-determining Y-chromosome. Apart from initiating sex determination, the Y-chromosome does little that is functional, so any impact on gender-typed behavior would have to be quite indirect and therefore indeterminate.

If complex behavior cannot be genetically determined, then the entire tactic of using ancestral living conditions as a lodestone for identifying psychological adaptations is mistaken. Contrary to the views of early evolutionary psychologists, behavioral adaptations can occur on a very brief time scale. If moose become behaviorally adapted to new predator threats within a couple of generations, then human adaptive change must occur on a comparable time scale.

At such short time scales, genetic evolution is irrelevant given that this relies upon chance mutations that might crop up once in one hundred thousand years. Consequently, we must alter our definition of adaptations. These are no longer just gradual changes occurring over such slow time scales that are traced in the fossil record like the evolving leg bones of modern horses from tree-browsing ancestors.

Adaptation occurs on a much shorter time scale than previously imagined. Darwinism places differential reproduction at the heart of its explanatory mechanism. Whatever succeeds breeds. Yet, if adaptation occurs on a very short time scale of one or two generations, then the underlying mechanism cannot hinge on reproductive success. If reproduction is not central, then gene selection cannot be in play either. How does one define adaptation knowing what we now know about adaptive change occurring on a very brief time scale?

We live in an age of considerable change, unlike anything seen previously, and this rapid change cannot be due to gene selection. Contemporary changes are not random, however. To begin with, we respond in predictable ways to environmental variation. With universal literacy and the digital age, our brains are forced to process larger volumes of information in ever briefer times. This has the effect of increasing brain efficiency, or intelligence (the Flynn effect). It is a beneficial response to environmental change and is therefore adaptive.

The Flynn effect has a clear parallel with the improvement of learning capacity in experimental animals raised in an enriched environment. In other words, we are responding to environmental enrichment using mammalian brain flexibility that is a basic aspect of our biological heritage and involves sensitivity to varied environments early in development. So gene selection is far from irrelevant, after all.

Many changes in modern life are both highly predictable as a response to altered environments and also *inconsistent* with Darwinian expectations that we behave in ways that promote reproductive success. As health improves and more children survive, we experience declining fertility associated with contraceptive use, greater female labor participation, and urban life. Such changes are completely predictable from environmental alteration and can therefore be described as adaptive, even if they are not adaptations in terms of reproductive success. It is a good idea to study and understand such variation from an adaptive perspective rather than simply dismissing these phenomena as failures of adaptation, which is the conventional tactic in the social sciences.

If change occurs much too rapidly for gene selection to work, we are forced to look at other mechanisms facilitating adaptive change. These obviously include social learning that allows moose to get acquainted with their current predator threats or chimpanzee juveniles to acquire a style of fishing for termites that is distinctive of their local population.

There are many other adaptive mechanisms at play, most of which get cast in the shade by the very bright light shed on gene selection. In addition to adaptive development—as exemplified by the Flynn effect— these include the responses by individuals to the dynamics of their local mate market and the impact of health and nutrition upon work effort. A more comprehensive consideration of such mechanisms offers a better picture of how we are adapted to our current environment.

Scholars often object to the ambiguity of calling environmentally induced changes adaptive, yet such usage is fully justified. After all, an adaptation is a predictable response to environmental change that generally produces positive effects for the organism. Virtually all adaptive changes in phenotype are multi-causal, involving a complex developmental history predicated on many environmental and biochemical parameters. So nothing in biology is ever simple, and there is rarely a simple relationship between genetic change and phenotypic variation.

Heretical though it may seem, it is time to broaden the picture of adaptive change to include a variety of nongenetic methods of selection. In the modern world, these include the attainment of social and economic success in a world of increasingly differentiated status differences and wealth inequalities. These can have a profound impact on basic biology from stature to obesity and overall health and longevity.

Reproductive success is less of an issue in the modern world as high-status people no longer raise large families. Competition has not died, of course. Evidently, elites prefer to invest a great deal in a smaller number of offspring, thereby securing their hereditary privilege and long-term advantages in health and well-being. Cultural selection is commonly floated as an alternative to gene selection. I argue that this can be ruled out for preliterate societies and has minimal value as a scientific theory due to circularity among other basic problems.

The unifying concept of *Evolution in the Here and Now* is that societal variation may be understood in terms of a combination of individual and group adaptations to environmental conditions through mechanisms other than gene selection. In my opinion, this formula is the missing ingredient whose absence has impeded our understanding of our own species and its variation across societies. This approach finds a great deal of order that was previously hidden.

While good evolutionary biologists have always recognized a multiplicity of mechanisms through which organisms achieve adaptation to their environments, evolutionary psychologists overplayed genetic determinism from the beginning. It is time to correct this misstep.

It is not comfortable to think differently from the mainstream, but it is not difficult either. All it requires is a level of honesty and openness to one's own intuitions and to empirical evidence. What is really hard is to convince mainstream thinkers in either evolutionary psychology or cultural determinism that their particular approach is faulty. Why am I so sure that my approach is right and that the others are wrong? One answer involves intrinsic weaknesses of the theories themselves. The other—teased out in the course of this book—involves cross-societal research supporting adaptation to current environments.

Evolutionary psychology began as a reaction against the cultural determinist opinion that each society pursues its own random walk whereby accidental changes in thinking and technology propel other random events in a historical sequence that a skeptic described as "one damn thing after another." Every society is a distinct experience with its own knowledge, folklore, customs, traditions, fashions, languages, religions, gender conventions, conflicts, and political institutions. These may spread over time or across geography, but their propagation and progress are essentially random.

Evolutionary psychologists were very unhappy with the notion that societal differences are essentially random. To Darwinian biologists, such notions are scientific heresy. After all, Darwinian theory strips the complexity of all life down to two simple principles—survival and reproduction. Of these two organizing principles, reproduction is more important, and survival is just an organism's way of propagating more copies of itself, a phenomenon that modern Darwinians reduce to gene selection. Genetics was beyond Darwin's ken even though he was a contemporary of Gregor Mendel and likely possessed an unread copy of Mendel's paper on pea breeding experiments that discussed how parental characteristics get segregated in the offspring.

Instead of being overwhelmed by the spectacular variety of new life forms he witnessed on his *Beagle* voyage, Darwin looked for adaptive variation in it all, whether it was lizards making their way in a saltwater environment or finches changing their bill shape in response to varied food sources as they spread throughout the Galapagos archipelago illustrating clear signs of descent with modification from the mainland species.

In the Darwinian world, all is change. Fossil animals and plants are mostly very different from the contemporary species who came to replace them. Indeed, if we were to find a fossil bed from one hundred million years ago that contained species exactly the same as all modern ones, we would know that Darwinian evolution—whereby species evolve to exploit changes in their environments—was false.

Ironically, in their rebellion against cultural determinism, evolutionary psychologists highlighted cross-cultural *universals* that are the antithesis of adaptation and change. In hindsight, this was a mistake. We know that Darwinians wished to discredit the ethnographic representation of human societies as varying randomly around the globe. So they focused on evidence of commonalities across all societies and used these as evidence that our genetic heritage adapted us to the generic problems faced by hunter-gatherer ancestors.

Such hypothetical cross-societal universals focused heavily on gender differences in psychology, such as men's greater propensity to seek casual sex or their willingness to take risks. The problem with such claims is that these kinds of gender differences are unstable. Most are shrinking or disappearing as male and female economic specialization declines in modern societies. Women make up a substantial proportion of those using online pornography, and most have sex before marriage. Modern women also take more risks than their mothers and grandmothers as illustrated by greater participation in contact sports and greater risk taking in everyday life. These tendencies are borne out in traffic accidents, and teenage women are now as dangerous on the roads as teenage men.

While evolutionary psychologists wished to challenge the sheer chaos of cultural determinism, the notion of universals is a flawed Platonic concept that is inconsistent with adaptation. After all, if we cannot even claim that all moose are afraid of wolves, it is hardly credible to claim that all women are risk averse or less interested in sexual pleasure. In the real world, behavioral adaptation to the environment is a lot more flexible, faster, and a lot more subtle than the universals concept allows.

Adaptation is now—and always has been—concerned with the capacity for change to match the current ecological niche. Early evolutionists were inspired by the fossil record and studied long-term changes as preserved in fossils. As this became more complete, comparative anatomists could follow changes over evolutionary time measured in tens of millions of years. They could trace out long-term changes in anatomy. One venerable example concerns leg bone modifications as modern horses—who prospered as fast-moving prairie grazers—emerged from smaller leaf-browsing ancestors. Contrary to such glacial anatomical adaptation, behavior changes in very short time scales. Only a single generation was needed to adopt new nesting preferences for the previously endangered Mauritius kestrel that began nesting on safe cliffs rather than on trees that were vulnerable to predators. Israeli black rats expanded their range considerably by mastering skill in extracting the seeds from pinecones, also in a short time.

Any book on adaptation must consider the phenomenon of change as organisms become better adapted to available opportunities in their current environment. The speed and scope of change are particularly marked for humans. From early on, our species spread widely into new habitats, mastering the complex problems of subsistence and survival in radically different local ecologies having considerable climatic variation.

Initially, the demands of the local environment were the primary selective agent, but the acquisition of fire and other technologies gave our ancestors increasing ability to shape their own environments, sometimes in dramatically destructive ways, such as the Pleistocene Overkill, where the disappearance of large game animals soon after the arrival of human populations in the various regions of the globe suggests the use of sophisticated weapons used to kill from a distance.

However ingenious early humans might have been in exploiting the world's resources, they always reached some sort of equilibrium upon bumping up against the carrying capacity of the environment. So agriculture permitted growth of much more food than could be collected without cultivation, permitting a huge rise in global population. Even so, the human population was limited by food supply. English economist Thomas Malthus propagated the depressing perspective of a world where real improvement in the lives of the masses was an illusion. According to him, temporary success in producing more food simply inflated the population size to the point that more individuals starved, pushing down numbers to their inevitable limit. The only possibility of real improvement in the lives of workers came with catastrophic events, such as plagues and wars that shrank the population and left more resources available for each individual. Charles Darwin found in Malthus a key inspiration for his pivotal concept of natural selection and so did Alfred Russel Wallace, who also discovered evolution by natural selection.

Malthus's theory also influenced subsequent economists who sought to understand phenomena such as the rising cost of wages and increased living standards in Europe following the Black Death. Depressing or not, Malthus provided keen insights into how the world actually works. Yet his ideas apply only to societies where population is restricted by food supply. Malthus assumed that the amount of food available was a fixed quantity that would never change. The modern era with its adoption of industrial farming techniques permitted a huge acceleration in food production and consequently a global population explosion.

The big story in human ecology may be overconsumption of global resources, but the various technological revolutions have altered human beings as much as they changed global ecosystems. This is fairly obvious in respect to nutrition, health, longevity, and so forth, but the psychological impacts were equally profound. Just as improved physical infrastructure protects us from the threats posed by nature—by creating artificial microclimates in homes, cars, and places of work—innovations in information technology thrust us into a virtual world where many of our transactions are conducted on electronic devices rather than face-to-face. Of course, this becomes ever more obvious as the artificial (or "augmented") reality revolution plays out. When that comes to fruition, the distinction between virtual and real worlds may break down along with the distinction between artificial and human intelligence.

With increased involvement in electronic networks, our brains will receive more stimulation than ever before. This might have the effect of increasing intelligence, at least if one subscribes to the theory that the Flynn effect of rising IQ scores in developed countries is partly due to greater demands on the brain for rapid processing of information. As artificial intelligence greatly exceeds human abilities, however, it is possible that much of the hard work of information processing is relegated to intelligent machines and that our brains come to atrophy from diminished demands placed upon them, just as physical fitness declines thanks to the sedentary lifestyles of modern populations after machines replaced manual work.

Such phenomena pose major theoretical problems for adaptive evolutionary theories of human behavior. Simply stated, we may no longer be able to define the environment in the sense that Mbuti hunters are adapted to a forest ecology. Moreover, there is growing ambiguity about selective agents. This may no longer have much to do with reproductive success given that human fertility declines steeply in developed countries where infant and child mortality are close to zero, rendering gene selection weak or inoperative.

For Aristotle, writing in classical Athens, as for many other scholars before and after, the natural analogy for human societies was a beehive colony where the group prospered thanks to the cooperation of different castes, including the workers inside the nest, the soldiers defending the hive, foragers bringing back pollen and nectar, and the reproductives consisting of the queen and drones. It is hard to deny that bees and other social insects, particularly ants, offer a good analogy with the division of labor characteristic of all complex human societies. Yet there is one significant flaw. Bee societies likely remain much the same over millions of years. Their subsistence economy and division of labor never change substantially, although each hive does go through a developmental cycle with the queen initially having to fend for herself when founding a new colony before handing off every function apart from reproduction and producing a chemical that attracts workers to the hive. Yet it is always virtually the same cycle. By contrast, human societies change radically over time, as do the individuals within them.

Phenomena such as urbanization or progressive economic development have no perfect analogy among the social insects. The best parallel is the emergence of locust swarms from solitary individuals. The weakness of this analogy is that locusts quickly revert to the solitary form whereas complex human societies may get progressively larger. This flaw disappears if one accepts that all complex human societies are unstable as was true of previous civilizations.

Once our remote ancestors exited a forest-dwelling lifestyle, they wandered restlessly in search of new habitat. Other great apes live pretty much as they have always done (at least until human activities threatened them with extinction). Many scholars attribute human expansion to the development of tool technology that facilitated extraction of varied foods or killing of prey animals from a safe distance. Others emphasize rising intelligence and the propagation of social knowledge that helped our ancestors to master many new ecologies in their trek northward across Africa and into every other continent.

Whatever the secret of human success was, there is no doubt that we mastered far more different subsistence ecologies than any other primate—or any other mammal. No global ecosystem is safe from human intrusion and exploitation. Our species is adapted for change, and we see this not just in variation over long time periods but also on a time scale of a few generations. Models of evolutionary change that deal with very slow transitions, such as modification of horse leg bones to grassland life, are mostly irrelevant to human behavioral adaptation. It is like using a telescope when a microscope is needed. My intention in this volume is to supply the theoretical equivalent of a microscope that helps us to grasp the speed of human adaptation to varied environments.

Rapid adaptation to change is not a pan-primate phenomenon. Indeed, it is characteristic of all mammals, including moose, as we have seen. The same is true of invertebrates. One of the most startling cases involves locust swarms that develop from solitary animals. The precipitating event is rain in the desert that increases food availability and reproduction rate. The grasshoppers also experience a radical increase in sociability. Increased population density raises the level of social interaction with profound consequences for behavior. These include reduced territoriality, increased sociability, and a tendency to move together in swarms to the consternation of farmers whose crop fields lie in their path.

Grasshoppers offer a remarkable instance of adaptation to change in the food supply. Human civilizations can be interpreted in similar terms because there is a precipitating increase in food supply, rising fertility (or survival), and a large increase in population density with many subsequent changes in brain biology and social behavior.

Increases in human population used to be cyclical, but the present era of industrialized agriculture, aka the Green Revolution, sees a sustained population explosion and continuous increases in most forms of production, generating a prevalent increase in living standards around the globe. All developing economies undergo similar social changes from increased leisure spending and declines in gender specialization at work to improved education and increased intelligence.

Such generally favorable trends spring from increased productivity associated with technological development. Such change can last over multiple centuries, but we still do not know whether there is any stable equilibrium on the horizon. Could a global civilization live in harmony with planetary ecosystems, or does the current boom augur an even bigger bust?

From early in our origins as a species, human adaptation involved restless transition with many changes in location and habitat. Why did they keep migrating? That is a rather imponderable problem. Yet there are two logical answers that are not mutually exclusive. First, they moved because they could. They had the technology and smarts that permitted them to master new territory relatively quickly. Second, they moved because their current habitat had become untenable, whether due to climate change or their own activities.

The second option is suggested by the Pleistocene Overkill that followed human migration to most of the major regions on the globe some forty thousand years ago. Large grazing animals were no match for tools that could kill at a distance, particularly when combined with the power of intelligent strategies, such as stampeding herds of woolly mammoths over cliffs. No doubt there was a corresponding boom and bust in the human population as their large prey animals were hunted to extinction without any caution for the future.

Considered in those terms, surges in human population merit comparison with the opportunism of locust swarms. If early civilizations sometimes destroyed their ecology on a local scale, the current global civilization with planetary integration of supply chains and markets is doing the same on the largest possible scale, as many conservationists complain.

If the analogy between human civilizations and locust plagues seems too bleak, then everyone can agree (a) that the human condition lends itself to continuous change, and (b) that modern societies are characterized by more rapid change than ever seen previously. If we want to describe or understand our own evolutionary history, the last thing that we need is a Procrustean genetic determinism that claims we have not really changed under the skin for a million years.

The model of a small stable population genetically adapted to a generic hunter-gather way of life does not really work for our species. Of course, the myth of evolutionary stasis really falls apart with the Agricultural Revolution, when a huge increase in food production yielded a corresponding rise in population. Agriculture ushered in more complex status-graded societies and diverse alterations in human biology and health.

Cultivating food facilitated the boom-and-bust cycles of early civilizations that spiked for relatively brief periods in different regions around the world at locations that favored intense cultivation based upon irrigation and drainage. For its part, the Industrial Revolution facilitated a worldwide society that was subsequently bolstered by the digital revolution and Internet communications. If we want to understand how humans are adapting kinetically to modern environments, a static inherited-genome approach is a nonstarter because change must be the master theme. Change needs to be understood scientifically rather than accepting the arbitrary drift proposed by cultural determinists. What are the nongenetic mechanisms of change that adapt our species to modern environments? Read on to find out!

I

THE VITAL ALTERNATIVES

Good salespeople bamboozle their prospects with verbal tricks. One is known as the *vital alternatives*. An effective door-to-door vacuum cleaner salesperson used to bring at least two models with him that varied in quality and price. The objective was to get the "punter" involved in a discussion of their relative merits, complete with a practical demonstration. By focusing the discussion on the merits of Model A versus Model B, the salesperson had already made an implied sale. Now all he had to do was to help the customer decide on which of the vital alternatives to select. Why is this a trick? By ignoring the real decision—whether to buy a vacuum cleaner or not—the salesman introduces a certain inevitability to deciding among the two alternatives, in other words, buying a vacuum cleaner.

People are skilled at choosing between rival concrete possibilities. The ripe fruit is usually chosen over the unripe one. Close kin are favored relative to strangers. The quick payback beats out delayed rewards. Unfortunately, we are not nearly so good at weighing up abstract possibilities that are not physically present. When choosing from a menu at a restaurant, there is no "nothing for me" listing. People still can and do sometimes say, "Nothing for me, thanks!" but it is difficult (and uncomfortable).

"None of the above" is always an option, but it is rarely an attractive one. Humans are not constituted as an "opt-out" species. Instead, we grab onto the best options that we can, select the best friends, best mates, and best foods.

Our concrete selections may be fine-tuned, but the null choice is left to the philosophers. Few people will say to themselves: "Given the quality of what is available to me in this place and time, I would be better off without any friends, so will fire all of my current companions." We are, after all, a highly social species who acquire mastery over the task of cultivating friends that enrich our lives. Throughout our evolutionary history, those who selected even mediocre friends fared better than those who opted for a life of solitude. The same remains true today.

To be human, and even to be alive, involves making good choices. A female bird selects a good habitat in which to build her nest—one that has abundant food and security from predators—and, in the process, she obtains a vigorous mate who can defend this desirable home against competitors. It is because we evolved as choice-makers that the vital-alternatives trick works so beautifully for salespeople.

The vital-alternatives trick is alive and well in academic life as well. Whatever their peccadilloes, scholars are still people and as vulnerable to cognitive biases as other members of the species. In academic life, the vital alternatives often involve some sort of nature-versus-nurture distinction. Historians with a taste for the topic know that nature and nurture have wrestled through the decades. At some points, nature was on the mat and nurture prevailed. At other times, nature was perceived as everything. These vital alternatives alternate with the monotony of a rocking horse. It is only in the past two decades that we have begun to fully appreciate the extent to which nature and nurture work together to fashion individuals who may succeed in a specific environment.

THE HISTORICAL ROCKING HORSE OF NATURE VERSUS NURTURE

During the 1920s—a period of affluence and optimism—faith in the modifiability of human behavior was at its zenith. Behaviorist John B. Watson, who was deeply influenced by Ivan Pavlov's work on conditioned salivation in dogs, proposed a celebrated thought experiment. Give him a dozen healthy infants and he could turn them into any specialist— from physician to thief—if only he could control the environment in which they were raised. Such claims are easy to make but harder to back up with scientific evidence, as Watson himself acknowledged. We now know that differences in intelligence, which play such a big role in occupational specialization, are largely explained by genetic differences. So taking someone who is genetically predisposed for low intelligence and turning them into a doctor or a lawyer, occupations that call for above-average intelligence, would not be easy.

Luckily for Watson, this was just a thought experiment. There were no dozen children. Watson did not attempt to train them for some specialization. In other words, it was a bunch of hot air! The father of American behaviorism was as much a clever salesman for his behaviorist ideas as a serious research scientist. His natural exuberance landed him in trouble when he began a scandalous relationship with his lab assistant, Rosalie Rayner, that damaged his reputation in academic life, where he had been a leading luminary, editing the prestigious journal *Psychological Review* and conducting well-received research on brain development, conditioned fear, and other topics.

Watson subsequently applied principles of Pavlovian conditioning to marketing. He became one of the leading advertising gurus of his day and earned a lot more in business than he had ever got in academic life. Watson was the author of a controversial ad telling women that they could smoke with impunity if they used the right toothpaste. In this, and in other ads for personal care products, such as Pond's cream, he recognized that the real product being sold was sex appeal—a perspective that is part of the received wisdom of contemporary advertising but was unknown before Watson.

A decade after Watson's heyday in academic life, the pendulum shifted very much toward biological determinism, influenced by earlier ideas connected to intelligence testing and eugenics. On the one hand, there was the tradition of Henry Goddard's fine gradations of congenital feeble-mindedness that doomed their recipients to lives of misery, failure, and crime. (Goddard famously distinguished between morons, having an IQ of 51–70; imbeciles who scored from 26–50; and idiots who ranged from 0–25.) Then there was the tradition of eugenics—stemming from Darwin's cousin Francis Galton—that aimed to improve the gene pool by ensuring that such defectives would not be allowed to breed and cause degeneration of the population's genetics.

The eugenic perspective assumed a high degree of genetic determination of criminality, low IQ, and other problems. Of course, the Nazis implemented these ideas at the level of entire ethnic groups rather than just individuals. Proclaiming Germans as a master race, they set about forced sterilization of "inferior" groups such as Romanies. Tough though it might be to think about these issues today, Nazi ideas on genetics were widely shared in Europe and the United States. Indeed, leading figures in the eugenics movement hailed from the United States and England, both then global centers of scientific activity.

How can extremes of environmentalism and biological determinism succeed each other in such close historical proximity? Perhaps this vacillating extremism is born of deep ignorance and uncertainty. It is precisely because people knew so little about why individual differences emerge that they latched onto one extreme or the other with equally messianic fervor.

With Josef Mengele's grotesque experiments involving such Franken-
stein manipulations as injecting blue dye into the eyes of concentration
camp inmates, we can say without fear of contradiction that race science
was very poor science and actually not science but political propaganda.
After the war, as news of such atrocities leaked out, social scientists were
attracted to the opposite pole of environmentalism and bent over backward to show that group differences were often not due to genetic differences so much as environmental variation.

THE RISE OF CULTURAL DETERMINISM

The backlash against Nazi biological determinism was focused on eugenic ideas. American sociologists and anthropologists set out to discredit the
idea that some groups are inherently less intelligent than others. Their
main strategy was to highlight environmental influences on intelligence.
In the process, they strengthened the influence of cultural determinism in
academia and beyond.

The genetic determination of intelligence was a central conclusion of
Goddard's (1912) research on the Kallikak family, in which the inheritance of low IQ ("feeble-mindedness") was demonstrated over several
generations.[1] For Goddard and his contemporaries, "feeble-mindedness"
implied not just low IQ but also sexual promiscuity and criminality. Goddard later backed away from this research—agreeing with critics that it
was methodologically questionable—and softened his conclusions about
genetic determination of intelligence. Nevertheless, these ideas influenced the Nazis in their forced sterilization of Romanies and other groups
and individuals perceived to have feeble-minded or criminal tendencies.

Goddard's book must have been foremost in the mind of American
Justice Oliver Wendell Homes when he ordered the sterilization of a
young woman using the punchline, "Three generations of imbeciles are
enough." The young woman, Carrie Buck, had been committed to the
Virginia State Colony for Epileptics and Feeble Minded after she had
become pregnant due to rape by a relative. By committing her, the family
had hoped to protect their reputation. In reality, she seems to have been of
normal intelligence, as was her daughter, taking the air out of Justice
Holmes's punchline in the infamous *Buck v. Bell* Supreme Court case that
legalized state sterilization of those deemed unfit to hold down a job and
support themselves. Only Carrie's mother had low intelligence.

Energized by such eugenic atrocities, in addition to the "race science"
of Nazi Germany, American sociologists conducted research that brought
out a much more nuanced picture of the development of differences in

intelligence. On average, immigrant groups tested at Ellis Island by God-dard and his team scored lower if they were from southern Europe than if they were from the North, and Africans scored lower than southern Euro-peans. Of course, Goddard was eager to plug these data into a racist biological-determinist agenda and did not consider alternative explana-tions—a cautious approach that is always the hallmark of good science.

One obvious alternative factor is economic development and the fact that people from more developed countries scored higher on IQ tests because they were better educated and lived in more complex societies. Another obvious explanation of group differences in intelligence test scores would be group differences in nutrition. This is not a trivial factor because infants who are malnourished in the womb are born with reduced potential for intellectual capacity. Malnourished children also find it more difficult to pay attention in school, and school lunch programs were de-vised to correct this problem in America, an ostensibly affluent country where food insecurity remains a significant problem among people in the bottom quartile of income.

In America, as elsewhere, the eugenics issue was very much tied up with racism, as reflected in Goddard's fine gradations among northern and southern Europeans. Such distinctions are now known to be baseless, but they seemed more convincing to scholars and the general public in Goddard's heyday. The essence of racism is a conviction that by virtue of their ancestry and ethnicity, some groups cannot achieve at the same level as the dominant group. Once again, this seemed self-evident to many in Goddard's day in a world where the cards were heavily stacked against African Americans and others, including the Irish and Italians. Remember the "No Irish Need Apply" signs of the 1930s.

If sociologists wanted to prove that the poor performance of African Americans on IQ tests was of environmental origin, they needed evidence that if descendants of African slaves had the benefits of a middle-class upbringing, they could achieve at the same level as Europeans. This evidence was collected by studying the intelligence of black children adopted into white homes. Consistent with the environmentalist position, black children raised in white homes scored at the same IQ level as white children in what then seemed a stunning refutation of the racist version of biological determinism.[2] Sociologists devoted themselves to explaining all kinds of minority social problems in terms of discrimination—an exer-cise of assigning blame that has done no one much good. A more con-structive approach is to discover how minorities can be helped to succeed. With economic success, discrimination mostly takes care of itself (al-though one might point to historical exceptions such as the unique situa-tion of the Jews in Europe who derived wealth from banking that was denied to Christians based on religious usury laws).

For their part, psychologists devoted themselves to understanding key features of the environment that affected intelligence and academic success. They devised Head Start programs that offered intellectual stimulation for children in impoverished homes as an offshoot of President Johnson's War on Poverty. These programs initially produced very promising results with a substantial increase in IQ scores. Unfortunately, these benefits mostly vanished as children aged. This turned out to be a rather humbling intrusion of genetic determinism. As children get older, environmental influences on intelligence recede and genetic influences increase. No one really understands why this happens, but a couple of glosses are possible. The first is that many genes are expressed in age-dependent ways. For instance, the genes predisposing to Alzheimer's disease are typically expressed in old age. The second is that children are most dependent on their parents early in their lives so that they pay most attention to the parents themselves at this age. Anyone who tries to engage a moody teenager in conversation will understand what this means.

While psychologists and sociologists studied group differences in IQ from the perspective of environmental influences, anthropologists did ethnographic research under an extreme environmentalist agenda, known as cultural determinism. Cultural determinism was formalized by anthropologist Franz Boas and like-minded thinkers in other disciplines, including Emile Durkheim in sociology.[3] It assumes that humans are molded by the idiosyncratic differences among societies around the world.

Boas's student, Albert Kroeber, followed Durkheim in viewing culture as a sort of superorganism, like a beehive, where the whole is more than the sum of its parts and where the individual recedes in importance. He believed that cultural ideas have a life of their own independent of what happens in individual brains so that human societies operate in a sphere that is independent of biology and the natural sciences. In short, he was staking out territory for the new discipline of anthropology.

Boas believed that racial differences in intelligence and achievements were explainable in terms of experiences and education instead of the genetic determinism common early in the twentieth century where some "races" were considered biologically superior to others. Boas was a physical anthropologist and studied group differences in traits such as head shape. He favored social explanations when the evidence supported them and was otherwise open to hereditarian differences. His student, Kroeber, always favored social explanations over biological ones, however, regardless of what the empirical evidence said. He developed the idea of culture as an entity in itself that was originally propounded by Emile Durkheim, the father of sociology in Europe. Durkheim had argued that social facts call for social explanations. This extreme version of cultural determinism is alive and well in sociology, although it is currently under attack by adaptationism, which interprets gender differences in the context of adaptive specializations by gender as opposed to differences in social power.

For Kroeber, culture was superorganic and had a life of its own independent of what took place inside the brains of individuals. Subsequent cultural determinists also denied a role for biology in human social behavior and promoted the autonomy of culture from individual minds.[4] Kroeber's separation of culture and biology was extreme, and he asserted not only that culture was independent of individuals but also that the individual could not be shaped by the superorganic, a notion that is entirely different from how culture is described in modern sociology. He believed that possession of culture was a sharp dividing line between humans and other species so that biology and evolution are irrelevant to our understanding of human societies (or what he referred to as "history").

The hereditarian perspective, which Kroeber so vehemently rejected, derived from Darwinism and more specifically from Herbert Spencer. Kroeber rejected Spencer's application of the notion of Darwinian struggle to competition between individuals and to conflict among societies. Spencer believed that some individuals are innately superior to others and that this superiority permits them to succeed in competition over wealth and status. Similarly, he thought some racial groups are biologically superior to others. Spencer's ideas about conflict between societies are familiar to us today—at least in stereotype—as the Nazi claim to being a master race who were destined to conquer the world. Although laughable today, social Darwinism was taken seriously in Boas's heyday, and Boas championed the idea that all ethnic groups are similar in intelligence and aptitudes with most differences having environmental explanations.

Cultural determinism was not a completely new way of thinking, being anticipated by Christian theology according to which there was a sharp divide between the physical and spiritual realms, and thus between humans endowed with a spirit and animals devoid of one. Strictly speaking, Kroeber was not a religious person, but the dualism implied in his cultural determinism suffers the same scientific pitfalls as religious dualism. The key problem is to explain how humans—endowed with culture (or an immortal soul)—arose from the same Darwinian process as other animals supposedly devoid of these higher attributes. There is really no satisfactory solution to this problem. At the most basic philosophical level, cultural determinism thus lacks plausibility.

Even so, it remains the dominant paradigm in the social sciences today, just as genetic determinism was in the 1930s. Equally important, perhaps, it is widely believed by politicians, journalists, and the general public. So one hears a lot of grave pronouncements such as "Wage inequality is due to a culture of gender discrimination" or "Increasing sexual violence in high schools is due to a culture of rape." A moment's reflection reveals that such explanations are circular: when journalists invoke culture in this way, they are generally out of their depth but unwilling to admit it.

Evolutionary psychologists offer an opposing view according to which we remain hunter-gatherers under the skin, being adapted to the evolutionary past. This implies that our psychology is genetically determined to some degree (or it could not be genetically selected). Evolutionary psychology thus sees us as out of place in the modern world that we shaped for ourselves through modern technology, science, communications, medicine, transportation, agriculture, and commerce, to name a few of the most influential factors.

Both sociology and evolutionary psychology are obviously incomplete, and each ignores some important ingredients of what we are. Fortunately, there is a third alternative that unites our evolutionary origins with specifically modern influences. This is the perspective that humans, like all other species on earth, are adapted to the *current* environment, not some ill-defined ancestral one.

Contemporary adaptationism differs from evolutionary psychology mainly by requiring that more attention is paid to nongenetic mechanisms of adaptation, such as social learning, maturing differently in varied environments (flexible development), and responding differently to different social environments. Some of this adaptation takes place through changes at the individual level. It is also reasonable to discuss group-level adaptations as illustrated by marriage systems, technology, learned communications systems, warfare, and the formation of political alliances. I argue that modern societies are adapted to current conditions. That is a key to societal differences that has so far been concealed by the vital alternatives of cultural determinism and genetic determinism.

Margaret Mead was an influential proponent of cultural determinism in anthropology. She argued that sexual jealousy is socially constructed and absent in some societies. This improbable claim came back to bite anthropology and was a critical point of departure for evolutionary psychology in the 1980s.

Relationship jealousy in the general sense is likely present in all societies. Children value the vital connection with their mothers and react strongly to anything that threatens it, for example. Sexual jealousy is a lot more unpredictable. Margaret Mead even claimed that in the Island of Samoa, people were so laid-back about sexuality that jealousy did not exist.[5] As a naive young fieldworker, Mead was hoodwinked by her informants, who proffered all sorts of fanciful couplings for their own amusement.[6] Far from being the haven of sexual freedom that Mead imagined, Samoa had the same sexual conflicts and violent jealousies as other societies. All was not smooth sailing in the Samoan sexual world. Indeed, forcible rape turned out to be rather common, indicating that one primal conflict over sexual access was prevalent.

Margaret Mead's mistaken conclusions fell on sexually restrictive 1950s America like manna from heaven, offering a vision of places where there was more freedom of sexual expression. Mead's other main impact was to suggest that differences among societies in sexuality, or anything else, are largely arbitrary. She offered New Guinea, where sexual relationships were difficult and strained, as a counterexample to Samoa. Her cultural determinist approach was highly influential within anthropology. It also offered a perfect foil for the new approach of evolutionary psychology.

THE BIRTH OF EVOLUTIONARY PSYCHOLOGY

Long before Margaret Mead's deficiencies were exposed, her ideas of radical cultural determinism were challenged by anthropologist Donald Symons in *The Evolution of Human Sexuality*.[7] Symons's main theme was that human sexuality is very difficult to understand outside of an evolutionary framework. He pointed out that since men invest less in offspring than women, they have more to gain from casual sexual relationships, hence men's greater interest in physical attractiveness and pornography—evidence of wanting sexual experiences primarily for physical gratification. Symons concluded that women are more interested in romance novels than pornography because these depict the sort of high-investing men that they perceive as good husband material given that women compete with each other over male parental investment in their offspring.

Symons argued that human sexuality evinces many signs of Darwinian adaptation that are found in every society contrary to the cultural determinist perspective. For instance, men are most attracted to women who are not just highly fertile but at the *beginning* of their period of maximum fertility that extends from about ages twenty to thirty-five years. He saw this as a fine-grained masculine adaptation for tying up a prospective spouse's entire reproductive capacity. He omitted the underlying mechanisms in genetics and brain biology.

Does natural selection fine-tune masculine perceptual capacities in ways that maximize reproductive success? Perhaps men are attracted to women in their early twenties purely because the women are in their biological prime and putting out the strongest signals of health and fertility. After all, that is the age of typical beauty contest winners whether evaluated by male or female judges.

There is little concrete evidence for genetic influences on human mate selection, but the idea is not as improbable as it might seem. Early brain biology does affect mate selection. Thus, women who were exposed to heavy doses of sex hormones in pregnancy (via mistaken use of the birth control medicine Stilbesterol) experienced some masculinization of the brain and were more likely to be attracted to other women and report as lesbians.

Symons and other evolutionary scientists argued that human sexuality looks alike in most societies. This phenomenon provides evidence of adaptation (through whatever mechanism, genetic or nongenetic). It challenges the blank-slate views of Margaret Mead's cultural determinism.

Another key challenge to that extreme environmentalism came from behavioral genetics. Twin studies found that many personality differences were strongly affected by genes. For instance, more than half of individual differences in IQ scores are attributable to genetic ancestry. Such phenomena challenge cultural determinists because they suggest that human talent and accomplishments are not entirely the product of environmental influences. This is why Stalin saw genetics as such a threat to the egalitarian claims of Communism and put Mendelian geneticists out of business by the simple, if drastic, expedient of murdering them.

Contemporary students are presented with the following vital alternatives. According to cultural determinism, children acquire the ideas and habits of their society that descend to them via random drift across generations. They are not adapted to their contemporary environment. According to evolutionary psychologists, human behavior *is* adapted to the environment. However, it is not the *current* environment but the environment of evolutionary adaptedness to which we are matched thanks to genetic adaptations crafted by hundreds of thousands of years of life as hunter-gatherers. Stated like this, one would be excused for wanting a third option. That is what this book aims to supply. The third option is adaptation to current environments that includes genetic and environmental influences together.

THE THIRD WAY

Here is how the vital-alternatives trick works. By presenting a choice between two actual vacuum cleaners, the salesperson deflects attention from potential third options, including neither of the above. Nature and nurture emerged throughout the history of the social sciences as mutually exclusive alternatives. The third option includes *both* genetics and environment that are always required for adaptation to occur to the *contemporary* environment (which is the only one that really matters for living creatures). The third way involves an interaction of biology with experience. This is at the heart of adaptation to the environment, even in cases that were often mistakenly assigned solely to the nature compartment. This phenomenon is illustrated by how animals learn to be afraid of their local predators, something that evolutionary psychologists interpreted as genetically determined.

Behavior must be calibrated in adaptive ways to varied environments, and this works best if genetic and environmental components are united. Fear responses develop predictably in all prey animals, which is why it is easy to exaggerate the role of genetic inheritance. Prey animals go through predictable routines of paying attention to assess the threat, becoming alarmed, often emitting alarm calls, and fleeing for cover.

Although prey animals have a well-developed fear response in all environments, they cannot know in advance of what to be afraid. This point was highlighted by the reintroduction of wolves to wilderness areas where they had become extinct. Astonishingly, moose, who are their natural prey, had lost their fear of wolves and were said to be "naïve."[8] Adult moose soon learned to fear the wolves, however, due to being stalked and attacked, and passed on this fear to young via social learning. In other words, when a calf sensed that its mother was afraid, it became fearful and learned to associate that fear with the scent of wolves.

So fear of predators is not built in at birth as evolutionary psychologists had long assumed (including human fears of snakes and spiders). Either this is impossible (as seems likely based on what is now known about brain development) or it is less effective than combining genetic inheritance with individual learning. Fear of predators is not unique, and the whole notion of genetically determined adaptive behavior is problematic. Even the species-typical contact calling of herring gull chicks requires auditory experience for its development according to experiments by Gilbert Gottlieb.[9] According to psychologist Mark Blumberg,

> The so-called maternal instinct of female rats to retrieve a straying pup can be eliminated by rearing the mother with powdered food, thereby depriving her of the normal experience of handling objects (such as food pellets). Even more bizarre is the finding that profound fear of snakes exhibited by Japanese macaques, considered an instinct in monkeys as well as humans, relies on the monkeys having gained early developmental experience handling and eating live insects.[10]

Behavioral adaptation always requires both genetics and environmental experiences.

Scientists have only begun to grapple with the tough job of understanding just how adaptation is accomplished, whether at the individual level or at the level of social groups. Needless to say, with a more sophisticated understanding of how adaptation is achieved, few serious people can any longer be satisfied with the vital alternatives of social learning and biology. If we are asked why moose are afraid of their natural predators, it is no longer respectable to say, "Oh, it is instinct." Even the more knowing answer, "Oh, it is animal culture," is not very helpful without a thorough grounding in the evolution of antipredator tactics of prey species.

Adaptation to the environment requires appropriate functioning of brains and bodies that is strongly affected by gene selection, but it also calls for behavioral tweaking, such as the moose having a propensity to feel afraid in every environment but requiring individual experiences to target that fear at specific local threats. Such tweaking of adaptive behavior is likely the rule for most complex animals, including humans. In our case, adaptive variation even occurs at the societal level. So the practice of polygamy is more affected by proximity to the equator (where food plants produce collectable food throughout the year) than it is by religion. Of course, food abundance also determines polygamy among birds. The many mechanisms by which animals become adapted to their current environments are worthy of detailed attention.

2

MECHANISMS OF CHANGE

Darwin's brilliant insight into evolutionary change is one of the most celebrated ideas in science—and justly so. More than anything else, it is attractive because it is beautiful. The beauty lies in its simplicity and the ease with which it reaches out to embrace much of the great complexity we see in natural phenomena around us. At the end of the *Origin of Species*, he indulged in a rare moment of self-congratulation to reflect that "there is a certain grandeur in this view of life."[1] I have always been and continue to be a Darwinian, and it is disappointing that many evolutionary psychologists resist extending Darwin's insight into other mechanisms of change than genetic determination (which was, ironically, unknown to Darwin himself).

Although his personal humility might suggest otherwise, Darwin was a member of the elite who had married money in the form of the Wedgwood pottery fortune. He saw little gainful employment throughout his life, partly thanks to persistent gastrointestinal health problems that made it difficult for him to work full time. So he spent most of his life on a farm near London leading the life of a country gentleman. Darwin was fortunate to live in a place and time having great interest in science. Such fascination gave him the opportunity to undertake the *Beagle* voyage as a naturalist, this being one of a series of government-funded voyages of discovery. Sponsors of the voyage may have been more interested in extending the colonies than they were in advancing science, but that did not minimize the scientific opportunity placed before the young Darwin. He was little more than an amateur enthusiast but brought the essential attributes of curiosity, impressionability, keen observation, and a facility for detailed record keeping both in field notes and sketches.

Back in England, he could travel the globe again in his keen memory. Moreover, the success of the *Beagle* mission launched him into the scientific forefront. Consequently, he built up a network of friends who included the top biologists of his day, including Thomas Huxley, who would become known as Darwin's bulldog for his pugnacious defense of Darwin's work after the *Origin of Species* had been published.[2]

Long before that, Darwin acquired some fame as the *Beagle* naturalist, and a book based on his voyage journals became an instant best seller. He began a correspondence with other naturalists around the globe that continued throughout his life. Eventually, the mail became so voluminous that he could not hope to read everything. This might explain why Darwin was unaware of early experiments in genetic inheritance because Mendel's paper lay unread in his mail backlog. This is a shame because Mendel's work could fill a key hole in Darwin's theory up to that point, namely the mechanism through which parents propagate their traits to offspring.

The central question that Darwin sought to answer was that of adaptation. Why are species so exquisitely matched to their way of life? As so often in science, why questions are eventually reduced to how questions, those that relate to the causal mechanisms of this chapter. His station in life as a gentleman farmer equipped Darwin to answer this question in a remarkably simple way.

Darwin pursued many hobbies that were seemingly quite unrelated to his interest in adaptation. In this technical research, he studied the anatomy of plants and barnacles so as to unearth similarities and divergences among related species. As a gentlemen farmer, he participated in the popular craze for animal breeding and kept a pigeon loft. Pigeon fanciers used selective breeding to produce unusual feather patterns. They produced exotic-looking strains, such as pygmy pouters with long neck feathers and fantails with elongated tail feathers that they displayed after the fashion of peacocks.

From these amateur experiments, Darwin gleaned some of the key planks of his theory of evolution by natural selection. The first is that organisms are variable with some of this variability passing from generation to generation so that traits may be exaggerated by repeatedly interbreeding individuals who are high on the trait. For a pigeon fancier, the trait selected might be a longer tail or longer neck feathers; the traits are chosen arbitrarily by the breeder with the proviso that they must be heritable and also that they should produce an interesting or pleasing alteration in the appearance of an ordinary pigeon.

Knowing what he did as a gentleman farmer, Darwin came to a simple but startling intuition. What if the natural world acted like a selective breeder? Of course, natural selection is not arbitrary. On the contrary, it equips animals with exactly those traits that help them to survive, and hence to reproduce, in whatever habitat they make their living. For giraffes, having an extra-long neck—something of an engineering marvel of the animal kingdom—is an advantage because it allows them to reach high into trees and always find food of a leafy variety even when there is a drought that kills most other vegetation and forces grazing animals to move on.

NATURE AS THE BREEDER

Darwin deduced that there is a real struggle for survival going on for all species, humans included. A persuasive case for this thesis had been made by English economist Thomas Malthus, an author whose ideas on population influenced Darwin deeply. Malthus argued that human population was tightly regulated by the availability of food.[3] Add more food to the equation through a series of good harvests and the population rises, only to be cut back by starvation in the leaner years that are sure to follow.

If resources are limited, some individuals will be better suited to obtaining them than others are. They will produce more offspring. Given that offspring resemble parents, these will express the parents' favorable traits. So natural selection breeds animals (and people) that are good at surviving and reproducing. These ideas are powerful because they provide a causal mechanism through which every organism gets ideally matched to its habitat. It is equally good at explaining why giraffes have long necks; why scimitar hummingbirds have long, curved bills that match the calyx of flowers on which they feed; or any other of the exotic examples of apparent design in the animal kingdom.

Incidentally, Darwin was keenly aware of such adaptations because they were the stock in trade of natural theologists, such as William Paley, by whom he was deeply influenced in his youth. Paley argued that the beautiful match between animals and their way of life was evidence for a benevolent creator who designed his creatures to fit their habitat. Paley's perspective is the now-familiar argument from design. In a favorite metaphor, we are asked to think about a man finding a watch and inquiring where it had come from.

Obviously, if there is a watch, with its exquisitely engineered springs, escapements, and cogs, it exists thanks to the labor of some watchmaker. Of course, this is a spurious argument because it is circular. We already knew that watches are man-made, so concluding that a particular watch had to be manufactured tells us nothing new. Moreover, any conclusion about man-made objects has zero relevance for living organisms that are not man-made. Such deductive errors can be surprisingly compelling, a fact that casts doubt on the robustness of human reasoning. Although logically flawed, such arguments are emotionally satisfying, and the emotion often wins out.

If natural theology used nature as supportive evidence for theological claims, Darwin's theory of evolution did the opposite. If natural selection could explain why animals are ideally matched to their ecological circumstances, then it removed the benevolent creator from the picture. In other words, the "why" of theology was taken over by the "how" of natural selection. With the mechanism of natural selection in place, there was no need for a creator.

Darwin quickly realized the theological implications of his theory, and he found them disturbing. If evolution by natural selection was the true reason that organisms are adapted to their habitats, then the notion of a biblical creation would be fanciful. Evolved species were mechanically produced through the operation of natural selection over countless generations and required no divine intervention, no watchmaker, no benevolent designer.

Darwin knew that a peaceful coexistence between the biblical Genesis and his theory of evolution by natural selection was not possible. That conclusion was immediately apparent to him as it would be to anyone else who read his work. Publication would unleash a furor where organized religions lined up against the new heresy. Authors of such explosive theories are not always treated well, and Galileo provides a good example of how Catholicism, in particular, treats heretical scientists. Galileo escaped execution by the skin of his teeth and spent his life under house arrest. The Protestant Church of England was also far from broadminded.

Darwin's wife, Emma, was a deeply pious woman who rejected the application of evolutionary theory to humans, or any other threat to her religious faith, thereby opening up a chasm between them. This tension may have kept Darwin on the fence so far as publishing his theory was concerned, something he would likely have done sooner had he been single. Maybe he pretended to greater agnosticism than he felt for the good of his marriage, which needed to be maintained for the sake of his beloved children.

Given these many personal insecurities, it is understandable that Darwin would have waited two decades until most of his children were raised. There is another plausible reason that he delayed publication until a rival, Alfred Russell Wallace, threatened to beat him to the punch. At this juncture, to have delayed would merely have lost priority and would have been pointless if the cat was already out of the bag. An additional reason is understandable academic caution.

A key component of Darwin's theory was the idea that species may change as they adapt to varied habitats. Of course, this was controversial for theological reasons and religious people, including Emma Darwin, believed that animals were created in their current form and did not change. One way of demonstrating "transmutation" of species was to construct a family tree of related species showing descent with modification from a common ancestor. For instance, comparisons of the anatomy of humans and great apes meant that we could only have descended from a common ancestor and the resemblance was compelling when one compared immature apes to humans. Another relevant piece of evidence came from the study of fossil remains of animals from different geological periods. Anyone who sees a fossil woolly mammoth is struck by its resemblance to contemporary elephants, suggesting that they must derive from a common ancestor.

One of Darwin's great passions at the time was the study of marine invertebrates, specifically barnacles, a small shellfish that adheres to ship timbers. Darwin devoted eight years of his life to intensive study of the anatomy of living barnacles of different species. His extensive work on barnacle systematics was well received by the scientific establishment, cementing his reputation as a research biologist. Such detailed and time-consuming work is what cautious scientists typically do, and one can infer that having performed such spadework, Darwin was probably accumulating evidence that he knew would be useful for defending his theory when it was finally published. In any case, he found that related species could be arranged into a family tree that supported transmutation of species and falsified the notion of a fixed creation.

As to the nuts and bolts of natural selection, Darwin relied heavily on Malthus to establish that there was a struggle for existence, at least so far as humans are concerned. Yet there were profound technical holes in the evidence for evolution by natural selection. Darwin assumed quite reasonably that there was natural variation between individuals, upon which natural selection might act. If the theory were correct and the most fit individuals were selected, then natural variation would be expected to fade over time until it almost disappeared. For example, all giraffes today develop long necks. Assuming that their ancestors had the shorter necks typical of other ruminants, how would natural selection have preferred long necks if there were none? This is not an easy riddle to solve, and asserting that selection works on the most gradual of changes among individuals (as we now believe it actually does) was problematic.

Even if neck length was heritable, how could one be sure that the minute changes upon which selection acted were themselves heritable?

After all, they might be incidental effects of better nourishment or straining to reach high branches, as Lamarckians supposed. Darwin's colleagues knew nothing about genetics or mutations. To the question "Why do offspring resemble parents?" the only available response was "Because they do." That couldn't be shakier as a scientific premise.

Without knowing about genetic inheritance, evolution by natural selection was a bit of a leap in the dark. Of course, animal breeders knew that artificial selection worked, so there was good reason to believe that natural selection would work also. However, the mere fact that some arbitrary trait, such as neck-feather length of pigeons, was transmitted from parents to offspring did not mean that all traits relevant to differential survival and reproduction were similarly transmissible.

When one combines the personal costs that Darwin faced by publishing a heretical account of the origin of species (including humans) with his monumental scientific caution and the many genuine gaps in Darwinian evolution, his hesitation is understandable. It is possible, even probable, that without hearing from Wallace, he would have postponed publication until near the end of his life.

As it was, he was goaded into publication only by the knowledge that he risked losing priority to his rival, Wallace. Wallace had read Malthus and drew similar conclusions from it, namely that there was a struggle for existence and that those individuals most favored for survival could propagate their characteristics into subsequent generations, thereby ensuring that each organism was well suited to the way that it made a living.

Having independently conceived of evolution by natural selection, the young naturalist wanted to run his idea by a leading luminary in the field of natural history and sent his summary for Darwin's perusal. The recipient must have felt that the sky was falling and was galvanized into action. In a model of scientific rectitude, both authors presented a joint paper to the Royal Society so that priority was shared. Darwin then went to work to ensure that evolution by natural selection became his legacy and that Wallace was relegated to a historical footnote. Of course, he had many advantages in this contest, including high status in the scientific community and a lifetime of careful observation that could be mined in his initial book-length treatment of evolution by natural selection (the *Origin*) and other ground-breaking contributions, such as the theory of sexual selection introduced in *The Descent of Man*. At the end of the day, Darwin received the lion's share of the credit, not just because he was entrenched in the scientific establishment, as represented by the Royal Society that had previously honored his systematics research, but also because he had close friendships with the most prominent English biologists of the time. Apart from these "who-you-know" advantages, Darwin received the credit for evolution because he worked harder on the theory than anyone else.

However hard he worked, Darwin's theory of evolution by natural selection had serious gaps and flaws in the mechanisms of inheritance and evolution. These received a great deal of criticism after the theory was published. Ironically, some key fixes were available in the breeding experiments of a monk, Gregor Mendel, who lived in what is now the Czech Republic. Mendel sent reports of his elegantly designed experiments to leading scientists of his day, only to be answered by a deafening silence. One recipient may have been Charles Darwin. Either Darwin never read the paper, or he failed to see the relevance of garden peas to his theory of evolution. Mendel was so disappointed by the nonreaction that he abandoned scientific work and died an early death. His papers were independently discovered by no fewer than three researchers in the same year at the turn of the twentieth century. Clearly, biologists were now prepared for a more rigorous treatment of the mechanisms of heredity that would cast a long shadow over behavioral sciences in addition to biological disciplines.

MENDEL'S CONTRIBUTION AND THE CAUSAL HOLES IN DARWIN'S THEORY

Darwin's theory could not explain why offspring resembled parents. Faced with universal ignorance on this topic, Darwin adopted the worst strategy possible. He fudged it, devising a fictitious account in terms of "pangenesis" that is reminiscent of Aristotle's musings on this subject. This theory was not only indefensible in itself; it yielded endless problems by needlessly drawing the reality of evolution into question. In science, honest ignorance always trumps pretended knowledge.

The basic idea was that all organs of the body give off "gemmules" that accumulate in the germ cells to be combined in the next generation through a blending of characteristics from both parents. This speculative explanation for resemblance between parents and children exposed Darwin to charges that evolution was not tenable. If parental traits get blended in each generation, directional selection is virtually impossible because the favorable traits of one parent get diluted by the characters of the other leading to an expectation of trait homogeneity over the long run. Of course, Mendel's research on garden peas supplied the perfect solution to this problem. Mendel established that inheritance was particulate, or, as we would say, "digital." This means that inheritance of a trait could be traced to one parent or the other, as illustrated by eye color. Characters did not have to blend. Some traits prevailed over others: peas with round seeds being dominant over wrinkled ones, for instance. This meant that offspring receiving the round genetic trait always produced round seeds. It was only when the wrinkled trait was received from both parents that the offspring produced wrinkled seeds.

Even those who were sympathetic to evolution by natural selection hated pangenesis as a cobbled-together tale that created more problems than it solved. Philosopher Michael Ruse offers this rather scathing review:

> There are many problems with Darwin's theory. The theory is dreadfully ad hoc, with all kinds of conceptual gaps. For instance, why are features sometimes transmitted in a nonblended way (let us speak of this as "particulate transmission")? Why does a child always have a whole penis, or none at all? Why do sexual organs fail to blend, whereas something like skin color does blend? Again, why do we sometimes get features skipping a generation? Presumably the gemmules can lie latent, but why? Darwin had no answers to these questions, nor did he have answers to obvious questions concerning the exact physical nature of the gemmules, and precisely how these particles end up in the sex organs. None of Darwin's contemporaries was very much enthralled by pangenesis and critics like Mivart (1871) really went to town. Today Pangenesis is on the scientific trash heap, along with phlogiston theory and Ptolemaic epicycles. [4]

Of course, Mendel's research plugged many of these holes. He concluded that genetic traits continued unchanged from generation to generation, even if they were not always expressed (as in the case of recessive traits). He demonstrated this by showing that recessive traits that were suppressed in the first generation were revealed in a second generation when the offspring received a recessive gene from both parents.

Evolution by natural selection has two essential ingredients: variation and selection. No reasonable person can doubt that selection works given that animal breeders do it all the time. If there is a struggle to survive and some individuals are better suited for success in their native habitat, their heritable traits will inevitably get selected out for exactly the same reason that artificial selection works.

Instead of the breeder playing God, so to speak, it is the rigors of survival and reproduction in nature that do the selection. As noted, the gemmule idea suggested that variation could not persist for long enough to be preferred by natural selection. Without variation to act upon, selection is impotent and evolution by natural selection is unworkable. For instance, infant mortality is so low in developed countries that any inherited resistance to diseases of the newborn cannot be selected. For similar reasons, it can be inferred that natural selection does little with modern human populations. This poses a problem for evolutionary psychologists whose struggle with refractory mechanisms resembles Darwin's problems. They have trouble marshaling the causal mechanisms needed to support their views.[5] Not all adaptations are inherited, however. Many arise in the lifetime of an individual. It is only by examining all of these mechanisms that we can grasp how each individual makes the most of their opportunities to thrive and to reproduce.

The real weakness of Darwin's pangenesis is revealed by his response to critics like St. George Jackson Mivart. He responded by resorting to Lamarckism, the notion that traits acquired by parents are more likely to be inherited. Lamarckism might seem like a weak theory because it lacks a plausible scientific mechanism and thus recapitulates Darwin's problems from pangenesis. Or at least that was true until the epigenetic revolution revealed how the genotype can be marked by parental experiences, thereby predisposing children to obesity, abusive parenting, psychopathology, and who knows what else. As it was, Mivart had a witty rejoinder. If Lamarckism really worked, why did Jews have to continue circumcising their male children generation after generation for thousands of years? Lamarckism also begs the question of the time scale of evolution. If Lamarck was correct in his claim that acquired traits are inherited, surely evolution could advance in a very few generations as opposed to the millions of years favored by Darwin based on fossil evidence.

PROBLEMS IN THE TIME SCALE OF EVOLUTION

Half a century ago, Lamarckism appeared to be dead. Now biologists are not so sure. It emerged that traits associated with parental experiences may alter descendants in unexpected ways. For example, if starvation is a feature of parental lives, offspring are more vulnerable to being overweight.[6] Analogous findings apply to abusive parenting, fearfulness, and probably many other traits of our species and others, as will be discussed later. The mechanism for such seemingly adaptive traits is even more shocking to (previous) received opinion in biology. It involves changes in gene expression.

It is difficult to understand how such changes could occur in a single generation and still get selected by an evolutionary mechanism. Yet organisms that are flexible in terms of a range of possible phenotypes likely have an advantage over those that manifest a single phenotype. Such "genetic polymorphisms" are common. In general, polymorphisms are triggered by some cue in the environment, such as increased predation or reduced food supply.

So a tiger salamander develops a larger body—suited for predatory cannibalism—when the population density increases.[7] Is this different from varying outcomes in epigenetics, such as abusive parenting affecting brain development and making the second generation more vulnerable to abusing their own children?[8] The defining aspect of epigenetic change is that it is heritable, but the two kinds of phenotypic change highlight mechanisms through which genetics underlies rapid adaptive phenotypic change. We know little about how such mechanisms play out in evolutionary time scales of millions of years. If tiger salamanders exist exclusively in a low-density habitat, will they eventually lose their cannibal polymorph? Similarly, if genes affecting parenting and impulsive behavior are marked by childhood stress, will they stay marked, or will the epigenetic change reverse itself? Clear answers to such questions are unavailable currently.

Understanding how phenotypes unfold speaks to how quickly an organism may react to an altered environment with adaptive phenotypic change. Leaving aside the issues of epigenetics and genetic polymorphism, it is worth asking how long it takes for natural selection to produce adaptive changes. Conventional examples of adaptation, such as the changed structure of horse legs as they adapted to life on grassy plains, suggest that many millions of years may be required for adaptive change to occur. Yet many adaptive changes in biology can be much more rapid.

Lactose tolerance is a feature of early human life when much of the dietary intake is from milk. Children everywhere can digest lactose because it is a critical component of breast milk. As humans age, lactose tolerance declines. There is one conspicuous exception. In most European countries, adults continue to consume dairy products, and lactose tolerance is retained. Of course, the availability of dairy products is due to the prevalence of dairy farming, a very recent occupation. Adult lactose tolerance is thus an adaptation to persistence of dairy products, such as milk, butter, and cheese, in the diet of adults. In countries where it is difficult to raise cows, adults have little dairy produce in their diet and lactose tolerance declines to the point that dairy products cause bloating and diarrhea.

Given that dairy farming has only been around for less than ten thousand years, the phenomenon of adult lactose tolerance among European populations demonstrates that evolutionary change can be much more rapid than the slow transformation of horse bones. There are several well-known reasons that evolutionary change can speed up. One is that selection pressure is intense. In a dairying society, one can imagine that people who digested milk well would be much more successful than those suffering severe lactose intolerance. Agriculture may have been a source of intense selection pressures for diverse reasons.

Humans were in close contact with domestic animals for the first time in human history and thus exposed to some of their diseases, such as cow pox and ringworm. As a result, humans have acquired immunity to some of these diseases. So our immunity to cow pox was used to create a vaccine against smallpox (a deadly illness) that protected humans against that epidemic scourge. We may never fully understand how such immunities developed in thousands of years rather than millions, but they likely involved intense selection pressures.

Another reason for speedy adaptation may be that the underlying biochemical mechanisms are already available or can be easily tweaked, whereas the legs of the horse likely depended upon many relevant mutations cropping up. Lactose tolerance is already present in infants, and its persistence into adulthood might be promoted by a single gene variant. Conversely, modifying the structure of a horse's leg is complex and likely required a large number of genes. Even though the fossil record preserves only the bones, there must have been extensive corresponding changes in musculature, innervation, vascularization, and motor control. Millions of years may have been necessary for the myriad underlying genetic changes to unfold.

Selection on other traits may be much faster. Ground finches from the Galapagos Islands, which were observed by Darwin, come in two types. Bills are either stout or fine. The stout bill is required to crack large seeds whereas the slender beak is better at manipulating small seeds. During normal years, both types of finch have little difficulty finding food and the proportion of both types in the population remains stable at close to parity. During drought years, there is a scarcity of fresh seeds and many of the fine-billed type die of hunger. However, the stout-billed variety have a major advantage because they can still find large seeds that are inaccessible to their gracile counterparts. Indeed, Weiner found that the proportion of stout-billed finches increased to around four-fifths of the population.[9] So adaptation took place in a single season.

Or did it? This phenomenon may be explained in terms of the sudden competitive advantage of the stout-billed variety. With diminished numbers of the other type, they experience a rise in numbers associated with the fall in numbers of their counterparts. Beak type is clearly determined by genotype, and the number of "stout-bill genes" rises due to natural selection in a single generation. Voila! Instant evolution!

There are two major problems with this interpretation. The first is that the genetic basis for stout versus fine beaks was already present. It did not arise *de nova* as a consequence of Darwinian evolution. If so, then it might be argued that nothing new was produced by natural selection here.

There was no new adaptation, so much as a playing out of selection on existing genetic variants. Of course, it might be objected that Darwin's theory referred to just such a playing out of selection among existing variability (that he could not explain).

Although the die-off of the slender-billed finches inevitably guarantees that the stout-billed genotype increases its representation in the population, this does not involve genetic novelty—no new mutation is promoted. Moreover, it is not clear that the change is permanent. During wet years when more small seeds are available, the slender-billed variety can make a comeback, and after many such good years, they might plausibly reach similar numbers to what they had before the drought. If that is true, then the proportion of each beak type could remain constant over large time periods measured in the hundreds of generations. If so, then what we are looking at is pure variability or noise. However dramatic such variability is, there may be no systematic change over time such as one might expect for gradual Darwinian change. We are witnessing random variation related to weather rather than directional selection. If directional selection were maintained, one beak variant would disappear. That this never happened suggests that populations having both variants are more resilient in the long run, and this helps explain why the polymorphism is maintained.

So we can be skeptical that gene-based evolutionary change can occur in a single generation due to natural selection. This does not mean that rapid evolution is impossible by other mechanisms, particularly behavioral ones. Without any gene selection, birds may change their nesting preferences in a single generation and rats may select novel foods and occupy vast new habitats thanks to different learning mechanisms. Rapid as they are, such changes are heritable and can be classified as evolutionary change, or adaptation, in my opinion. The crux is that they permit organisms to adjust to their habitat in new ways. The resulting stable match between the organism and its way of life satisfies the definition of an adaptation as an organism-habitat fit that improves chances of survival and reproductive success.

The bottom line is that the time scale of Darwinian evolution is not yet settled, which is why there is much that can be productively stated and investigated. If complex anatomical changes, like the adaptation of horse legs to life on grassland plains, take many millions of years, metabolic change and immunological adaptations can be much more rapid, judging from agriculture. Although settled agriculture has been around for little more than ten thousand years, there are a variety of adaptations to agricultural practices, including: adult lactose tolerance among dairying societies; eye color modifications by diet; immunity to animal diseases; and varying levels of alcohol tolerance, all of which may be interpreted as adaptive changes due to gene selection occurring within the past ten thousand years.

GENETIC ADAPTATIONS TO AGRICULTURE

Our remote ancestors likely had dark eyes, whether brown or black. Blue eyes are common in northern Europe and emerged only in the past six thousand years coincident with cereal farming there. This phenomenon shows that genetic evolution can be surprisingly fast.

In standard evolutionary theory, organisms are well matched to their way of life thanks to selection of favorable genes. Individuals carrying less favorable gene variants leave fewer offspring. When the way of life changes profoundly—as illustrated by an altered diet following the Agricultural Revolution—genetic adaptation gets to work and animals (including humans) become better matched to their altered way of life. This is what happens in the real world, and it stands in marked contrast to the evolutionary-psychology narrative of humans (uniquely?) failing to adapt quickly to environmental change. (We are presumed to be adapted to the environment of evolutionary adaptedness, meaning that we are better suited to conditions a million years ago compared to today.)

Agriculture enabled the production of large quantities of food, occasionally more than could be consumed. Some farmers used the excess to produce alcohol in large quantities. This was particularly true of rice-growing regions of Asia where rice wine was the favored alcoholic drink. Excessive alcohol intake was a problem then, as now, because alcoholics did less work and were less effective at providing for children. Natural selection may have favored the evolution of alcohol intolerance because those who were sickened by alcohol—and subsequently avoided it—left more surviving offspring. This could explain why so many people of Asian descent have a strong toxic reaction to alcoholic drinks that involves flushing of the skin.

Even light eye color (whether blue or green) is likely part of an adaptation to agricultural diet. [10] With the success of cereal farming in northern Europe, inhabitants experienced a deficiency in vitamin D in the diet. This vitamin plays an important role in immune function and health. In addition to dietary sources, it is also synthesized in the skin in the presence of sunlight.

One way of correcting for vitamin D deficiency is to reduce the amount of pigment (melanin) in the skin, thereby facilitating the passage of more ultraviolet light from the sun's rays. So natural selection for reduced skin pigmentation is a possible explanation for the paleness of skin of northern Europeans.[11] Reduced pigmentation also had the incidental effect of lightening eye color so that eyes are blue or green rather than the brown or black iris pigmentation that characterizes most of the world's population who produce more melanin that darkens the skin.

Contrary to the evolutionary-psychology narrative, there are numerous genetic adaptations to the Agricultural Revolution. These include genetic immunity to domesticated animal diseases—such as cow pox—as well as numerous adaptations to dietary changes. Such adaptations show that we are actually much better matched to the contemporary environment than many evolutionary psychologists assume. A faithful Darwinian account of human biology and behavior must bear these facts in mind. The bottom line is that the time scale of metabolic evolution is much more rapid than work in comparative anatomy might suggest.

Humans changed their location and their subsistence economy so often in the course of our evolution as a species that we had to be able to turn on a dime, so to speak. The pace of metabolic evolution is measured in thousands of years, rather than in millions, presumably based upon existing genetic variability rather than mutations. There are many other correlated changes in anatomy, physiology, and immunology. Human adaptations to agriculture are one link in a long chain of hominid adaptations to changing ecology and diet. We are now, and have always been, a species that moves geographically and changes behaviorally and biologically with exceptional rapidity. We are decidedly not the Paleolithic fossils presupposed by the theory of a fixed human nature.

HOMINID DIETARY ADAPTATIONS

Anthropologist Joseph Henrich argued that humans made such profound changes to their own environment and diet—for example by adopting settled agriculture—that this introduced strong gene selection pressures. Human bodies and brains have experienced rapid change in the relatively short evolutionary time since we diverged from chimpanzees around five million years ago. Our ape-like remote ancestors, Australopithecus, used stone tools to butcher meat some 3.4 million years ago, producing subsequent enhancement of their precision grip in terms of the anatomy of hand bones.

Early humans (*Homo erectus*) were distinguished by upright walking—probably an adaptation to living in a hot grassland environment in contrast to the shady forests inhabited by most primates. They had lateralized brains with a preference for using their right hands in tool fabrication. Although their brains were still small compared to ours, lateralization suggests increased competition for brain processing capacity, implying that they were doing more with their brains than Australopithecus had. Numerous other anatomical changes occurred in response to a lifestyle of scavenging and hunting. Not only was there an increase in the complexity of the diet, compared to mainly herbivorous ape ancestors, but a reduction in dietary roughage as humans targeted high-energy foods and subjected them to greater processing, eventually cooking meat and vegetable food to make it more easily digestible.

Extensive food processing meant a more energy-dense diet with less need for chewing and digestion. So there was a marked reduction in the size of teeth and jaws and consequent facial shortening. Smaller food volume permitted a reduction in gut size and consequently freed up metabolic energy that fueled a larger brain that had increased to 800 cubic centimeters by 1.8 million years ago. So rapid adaptation has always been a feature of human evolutionary history. Long before agriculture, humans were altering their own adaptive landscape with marked consequences for bodily and behavioral adaptations.

These changes are much more extensive than anything seen for other apes (who traveled less from their sites of origin and experienced less ecological transition). As far as we know, all of the other great apes are living essentially the same lives in the same locations, with the same diet and social behavior as existed five million years ago.

In marked contrast, our ancestors were always adapted to a changing subsistence ecology and social circumstances. They changed in response to novel ecological challenges. This fact contradicts the evolutionary-psychology narrative that we are currently adapted to an archaic way of life. The evolution of a precision grip in the service of refining stone tools with fine edges is now well understood in terms of hand anatomy and brain lateralization, for example.

The recent evolution of complex societies based upon extensive cooperation and trade is less well understood. In addition to their other feats of adaptive change, ancestral human societies somehow acquired a level of complexity in urban communities that outdid the social insects. How that occurred remains a challenge for evolutionary scholars that we must be willing to accept. It is only by being more open to multiple sources and mechanisms of evolutionary change that we can begin to address this formidable problem.

We must, as scholars, embrace the same flexibility that defines our evolutionary trajectory. That process begins by accepting a great variety of nongenetic mechanisms through which evolutionary change occurs. Some of the most compelling evidence for that perspective comes from genetics itself, specifically the epigenetics revolution showing that how genes are expressed is very much affected by environmental influences.

Genes are not selected because they contain a blueprint for some phenotypic trait but because they have some impact on how phenotypes unfold during development. In other words, genes are just one player in a welter of environmental and biochemical influences affecting development.[12]

THE EPIGENETICS REVOLUTION

Humans are masters of speeding up adaptation by nongenetic means, such as making fires and wearing clothes to survive cold winters. Early clothes included animal skins, demonstrating behavioral adoption of the biological adaptations of other species.

Such technological solutions are passed from generation to generation via social learning, and in recent generations, we also greatly benefit from the built infrastructure of roads, electricity, communications, and buildings. Valuable inherited structures are common in the animal kingdom, including examples such as beaver lodges, termite mounds, and tree-cavity nests of red-cockaded woodpeckers, which illustrate nongenetic solutions to survival problems.[13]

Comparisons among different human societies reveal a great deal of adaptive flexibility in development. Children raised in complex societies score higher on IQ tests, implying that the greater cognitive processing demands of modern societies increase intelligence (i.e., the Flynn effect).[14]

Social organization also responds to environmental differences. Monogamy is more common in societies having longer winters whereas polygamous ones cluster close to the equator.[15] Similarly, atheism increases in developed countries where there is less fear of early and sudden death.

In each of these examples, change is far too rapid to permit any genetic evolution, but our behavior changes to help us fit in with a different way of life. Human evolution thus occurs without change in gene frequencies. Natural selection does not have genes as a selective intermediary and acts directly on the relevant behavior. For instance, indigenous people who cannot fit in with modern societies dwindle in numbers and collapse as viable communities. In other words, their way of life gets eliminated by natural selection, painful and regrettable though that may be.

Natural selection can operate through genes, or it can operate without them. A fascinating recent area of research suggests that the influence of genes is different in different environments. This, of course, is the epigenetics revolution. Such phenomena have been convincingly demonstrated and have obvious adaptive advantages. However, there is currently no convincing explanations of how such complex adaptations arose over the long haul of Darwinian selection.

Behavior must be calibrated in adaptive ways with varied environments, and this works best if genetic and environmental components are united. Fear responses develop predictably in all prey animals, which is why it is easy to exaggerate the role of genetic inheritance. These animals go through predictable routines of paying attention to assess the threat, becoming alarmed, possibly emitting alarm calls, and fleeing for cover.

Although prey animals have a well-developed fear response in all environments, they cannot know in advance of what to be afraid. This point was highlighted by the reintroduction of wolves to wilderness areas where they had become extinct. Astonishingly, moose, who are their natural prey, had lost their fear of wolves and were said to be "naïve."[16] Adult moose soon learned to fear the wolves, however, as a result of being stalked and attacked. Mothers passed on this fear to young via social learning. In other words, when a calf sensed that its mother was afraid, it became fearful and learned to associate that fear with the scent and sight of wolves.

When gene expression is altered by the environment, there can be marked behavioral consequences. This phenomenon is highlighted by experiments on rodents raised in a stressful environment. These animals propagated greater fearfulness into subsequent generations, not by changing gene frequencies but by altering gene expression.[17] Instead of being a source of consistency in varied environments, genes that are expressed in one environment may be silenced in another. Notably, the altered pattern of gene silencing evoked by a particular environment is propagated to offspring via inherited banding patterns of the genome.[18]

Children raised in stressful homes are significantly shorter in stature despite the fact that height is one of the most genetically heritable traits. Psychological stress inhibits normal growth and development, probably by changing the way that genes are expressed. So a psychological (i.e., environmental) influence on growth alters gene expression. Psychosocial dwarfism is thus epigenetic. The same is likely true of the Flynn effect of IQ scores increasing with economic development. One smoking gun here is the fact that children who witness violence as they grow up score lower on IQ tests, implying that severe stressors inhibit the development of cognitive ability, presumably by altering gene expression.[19] Each of these examples points to the varied ways that environmental stressors affect gene expression.

There are many other possible cases of gene expression proceeding differently depending upon the environment. For instance, some specialists in the development of autoimmune disorders, such as asthma, believe that these are related to the sterile conditions under which modern children are brought up. If so, then children in developed countries are more vulnerable compared to children in less-developed countries where there is more exposure to diseases and parasites.[20] Their developing immune systems do not receive enough of a challenge from pathogens and therefore develop abnormally. Immune reactions are triggered to nonthreatening stimuli such as heavy exercise (that may precipitate asthmatic attacks). It is difficult to test this theory out without experimentally exposing youngsters to potentially dangerous diseases, but it does help explain why asthma and allergies become more common in developed countries where children are raised in very clean homes that insulate them from exposure to diseases so that their immune systems have minimal opportunities to learn which organisms are a genuine threat and which may be safely tolerated.

Another possible example of epigenesis concerns the rise of short-sightedness in developed countries. In Malaysia, many children wear glasses due to myopia. This high level of contemporary vision deficiencies can only be environmental given that gene frequencies in the population could not have changed quickly enough to account for such a rapid increase in myopia. The best available explanation is that children spend a lot of time looking at books or screens from close distances.[21] Apparently, this abnormal amount of close-vision focus in childhood alters the way that the eyeball grows, making it more elongated and increasing the chance of becoming short-sighted. The strongest evidence supporting this theory is that for indigenous societies, such as the Inuit, where there were neither books nor TV, there was also a very low incidence of myopia.

Due to epigenesis, children develop quite differently in different environments. The genotype comes with tricks up its sleeve, so to speak. Development can go in different directions, and which genes get expressed is determined by relevant environmental cues. We see this phenomenon at play in relation to environmental effects on intelligence in different countries—influences that range from parental stimulation and stress to neurotoxins.

How well a person does on cognitive tests throughout their lives is affected by early experiences that impair normal brain development. These include poor maternal nutrition in pregnancy that reduces birth weight, exposure to infectious diseases, contact with environmental neurotoxins such as lead and manganese, and reduced verbal stimulation from caregivers.[22] The known effects of these experiences on IQ (and even on brain size) are most likely explainable in terms of altered gene expression throughout early development, although the relevant biological mechanisms have yet to be worked out in detail. We do not know whether these influences may be transmitted epigenetically to future generations, but the high heritability of IQ scores certainly leaves this as a plausible possibility, given that standard measures of heritability include epigenetic as well as genetic influences.

What is true of intelligence also applies to work motivation and productivity. Researchers now recognize that there is a strong and clear connection between nutrition and school performance, but nutritional inadequacy also saps the work motivation of adults—they need to take more days off and cannot work for long at a high level of intensity.[23] Early nutrition can also affect subsequent vulnerability to obesity and related illnesses, including heart disease.

This phenomenon first emerged in connection with brief periods of hunger, such as that in Holland associated with World War II. Mothers pregnant during the hunger produced offspring who manifested greater health problems compared to other cohorts.[24] Animal experiments revealed that maternal starvation increased the likelihood of obesity. Moreover, vulnerability to obesity was found in subsequent generations. Evidently, when the fetus is nutritionally challenged, it reacts by developing a more energy-efficient phenotype. This trait is also propagated to future offspring, indicating epigenetic transmission of this adaptive response.[25]

Perhaps low-income populations have greater vulnerability to obesity and related diseases because they experience food insecurity and short periods of starvation. Of course, this epigenetic mechanism is just one possible reason that poorer segments of the population have a greatly increased risk for disease and early mortality. Exposure to environmental toxins and infections early in life are likely causes. Abusive parenting and epigenetic vulnerability to stress, which both span generations, are other reasons that have received a lot of attention by researchers.

Given that evolution matches phenotypes to ecological niches in many different ways, it follows that many adaptations work through nongenetic mechanisms. The simplest case of this involves social learning, but there are many other kinds of adaptive response to environmental variation.

NONGENETIC MECHANISMS OF EVOLUTIONARY CHANGE

The moose may lose all fear of its biggest natural predator. This reminds us that however slowly some adaptations emerge in the fossil record, others are shaped within the lifetime of the individual. This, of course, is a very different emphasis compared to the evolutionary-psychology narrative that is replete with accounts of species-typical psychological predispositions and behaviors that are allegedly transmitted genetically from generation to generation over millions of years. There is no doubt that some behavioral tendencies are affected by genes, but the notion of brain solutions to adaptive problems being genetically transmitted in whole cloth, so to speak, is a tall tale. As of now, there is no convincing example of this occurring, and the reality of brain development is very different. Indeed, there is no known mechanism by which genes can predetermine psychology and behavior and a host of examples of relevant environmental flexibility.

We have known this for a long time, but behavioral biologists like Nobel laureate Konrad Lorenz focused on species-typical traits that they assumed to be inherited directly through the genotype. Their influence spread to early evolutionary psychologists, such as John Tooby and Leda Cosmides, who developed the Swiss-army-knife model of human behavioral evolution.[26] The brain was conceived as a compendium of specialized problem-solving tools that dealt with adaptational problems, such as finding a mate or evading predators. Of course, our ancestors obviously did solve such problems, or we could hardly have survived as a species. The brain can also become specialized for solving certain problems but not in the way that they imagined.

Neuroscientists find that the brain is changed due to acquiring specialized skills, from playing an instrument to learning one's way around the streets of a large city like London. London's cab drivers must memorize all of the city's streets so that they can navigate without any map, a strangely anachronistic skill in the age of GPS navigation.[27] Based on the requirements of an earlier era, taxi drivers needed to have an expansive knowledge of the street map so as to drive customers efficiently from one point to another—a specialized skill, known locally as "the knowledge." London's taxi drivers have unusually good spatial skills to begin with, but careful research revealed that their hippocampus became more developed with longer exposure to the task of navigating London's extensive maze of twisting streets.

Such evidence is not just an intriguing quirk of London life. It speaks volumes about how brains become adapted to what we must do to make a living. This might not be so theoretically important if the brain also happened to have "modules" for specialized skills that are present very early in life thanks to "genetic programs." Yet such mechanisms are a figment: they are impossible. The first impassable hurdle is the normal pathway of gene expression, which is complex, messy, and environmentally contingent. Developmental geneticists have discredited the whole idea of a genetic program that sits on high and exerts hierarchical control over brain development, much like a blueprint dictates the architecture of a building.[28] Such a program would necessarily be more complex than the process of development itself, which is conceptually impossible because

DNA simply does not have enough information capacity to do this. Even if it did, there is no known mechanism by which a genetic program could control all of the necessary details of brain development.

In reality, brain development is highly malleable and contingent on incoming information from the environment, whether this is intercellular stimuli or external sensory information. When we see striking patterns in the natural world, we are inclined to attribute order to an external source. For natural theologists, this was the creator who expressed benevolence by shaping the habitat to support each animal's needs. Of course, Darwinian thinking replaced the benevolent creator with the heartless indifference of natural selection. In that world, the organism changes to fit the habitat rather than vice versa.

Complex social phenomena, such as the schooling of fish or the group patterns traced out in the skies by huge flocks of starlings, also seem to be purposeful or directed by a leader. Yet they are not. The exquisite group choreography of schooling fish or synchronously flying bird flocks are not directed from on high by a choreographer.[29] Instead they are emergent phenomena built up from the collective responses of all the participants: the order is bottom up rather than top down. Each animal responds sensitively to the movement of its neighbors so as to preserve an appropriate distance and avoid collisions.

Brain development manifests many of the same principles. Each cell in the developing brain responds to what its neighbors are doing. In particular, it strengthens connections to other neurons that fire simultaneously. Cells that are not stimulated in this way tend to wither and die. This phenomenon is known as neural Darwinism. Just as everyday Darwinism explains the apparent design pointed to by natural theologians, so neural Darwinism helps explain the development of integrated functioning in the brain, something that had earlier seemed impossible without some sort of program to direct the proceedings, this metaphor being widely used by early evolutionary psychologists. Such metaphors can seem to cast light on a problem, but they are fraught with errors. Real brains are very different from computers, most especially in how they are built during development.

Even so, neuroanatomists are impressed by the orderly arrangement of cells in the brain's cortex (or outer layer) that is sometimes compared to a microchip's etched circuits. When we observe such exquisite order in the brain, it is tempting to imagine that there must be some template from which it is drawn, rather like the etching blueprint for a microcircuit. Yet there is no such blueprint. Researchers concluded that the orderly function of patches of cortex that are functionally interconnected with their neighbors is derived from intracellular communication. The cells are functional blank slates and learn what they will do principally from the patterns of stimulation that develop between them and their neighbors. [30]

What is true of neuronal development also applies to behavioral organization that is accomplished in terms of similar cellular mechanisms. Such complexity is highlighted by the species-typical call of herring gulls, long treated as a genetically determined trait, by analogy with other bird calls that are so homogeneous—at least to human ears—that they may be used to distinguish among species and study their evolutionary closeness. Yet the herring gull call is far from being a product of genetic heritage alone. It actually requires responsiveness to the gull's own sounds produced as it develops inside the egg. Gilbert Gottlieb demonstrated this a long time ago in experiments showing that birds that were deafened surgically prior to hatching failed to develop their species-typical call. [31]

It is no longer controversial to state that no adaptation is completely genetic—genes are never expressed in a vacuum and this fact is brought home emphatically in the epigenetic revolution, suggesting that a surprising number of genes are expressed differently in different environments thereby facilitating a high level of adaptive plasticity in response to environmental variation. If many adaptations are affected in this way by the environment, then it is also true that some behaviors are largely determined by environmental variation in the sense that behavioral variation has no explicit tie to genetic variation.

One good example concerns the formation of dietary preferences among rodents. These are socially transmitted: young rodents develop a preference for eating whatever their mothers eat. This transmission process likely helps them to avoid ingesting poisons, but it is far from perfect, and the local variation in dietary preferences by rats shows that these animals pass up many good nutritious foods in their environment simply because they cannot recognize them as food. [32] Given the risk of toxicity of new foods, it may be better to follow the dietary preferences of Mom, who, after all, survived to maturity.

Perfect or not, the dietary preferences of rats are a socially transmitted adaptation. It would be wrong to say that such learning is entirely separate from genetic inheritance. After all, rats eat due to a variety of physiological mechanisms, including a sense of taste, and the mother's taste preferences could well reflect her genotype. Nevertheless, the key source of behavioral variation is clearly environmental. Transmission of the food preferences is not dependent upon gene selection, that is, the conventional model of Darwinian adaptation to the environment. Yet they are adaptive because they solve the problem of distinguishing between good nutritious food and nonnutritious toxic substances.

Social learning is an effective method of transmitting dietary preferences across generations. After all, the young must be raised by their mothers if they are to survive, and all are thus inevitably exposed to her dietary choices. This system is fail-safe, and it likely exists because there is no viable alternative, particularly in the case of omnivorous rodents, like rats. A similar case can be made about other examples of socially transmitted behavior, whether it is tigers being trained by their mothers to hunt or young chimpanzees learning how to extract termites from their nest using specially prepared tools.

Social learning is an effective way of solving the problems of survival and reproduction, and its prevalence in the animal world is gaining proper recognition.[33] One interesting difference between social learning and other ways of transmitting behavior is that it produces considerable variability and exploits the possibility of new solutions to old problems. Rats inhabiting the banks of the Arno river, in Italy, mastered the practice of diving to the bottom of the river to retrieve shellfish, a dietary habit that is not found for any other population of the species.

Individual learning generally produces more uniform results. Rats that learn to press a lever for food reinforcement may initially bite the lever or exhibit emotional arousal. Once mastered, the task becomes easier, calmer, and more efficient, being executed with the least necessary effort to acquire the largest food reward in the shortest time possible.

Such instrumental learning is obviously Darwinian, in the sense that it solves practical problems and helps the individual to survive. It is often compared to natural selection for another reason, namely that adaptive behaviors are shaped by the environment with correct responses being reinforced and ineffectual ones getting suppressed. This makes learning an extremely powerful method for animals to master the key problems posed by their environment. In many cases, their solutions are quite similar so that the acquisition of behavior individually may be quite reliable in the absence of social inputs.

Of course, the transmission of mechanisms for experiencing pleasure and pain would be much the same for all members of the species. Members of a species generally experience stimuli that favor survival and reproduction as pleasurable. Experiences that pose a risk, such as pain, extreme temperatures, or predators, on the other hand, are avoided in another powerful branch of instrumental learning. There is a lot of controversy about how quickly natural selection can respond to changes in the environment, but there is no question that animals may become quickly adapted to changing conditions through different forms of learning.

LEARNING AND ADAPTATION IN
NATURAL ENVIRONMENTS

Some animals are polymorphic, having different phenotypes, whether morphological or behavioral. Which morph is more common can vary rapidly depending upon prevailing conditions. Tiger salamanders are a good example. The larger morph has bigger jaws and is specialized for cannibalism. If the population is dense, cannibal morphs of tiger salamanders are favored and quickly become more common.[34] Similarly, guppies in river pools become more or less bold depending upon the number of predators about, and this phenomenon is attributed to natural selection working rapidly thanks to extreme selection pressure.[35]

These examples suggest that many species have multiple strategies for dealing with varied environments that are a predictable feature of their way of life. Over the long haul, each of these multiple solutions will be preserved. For if one adaptation was clearly superior to the other, then it would be the only one around today.

Adjustment to varied environments need not be either genetic or morphological: it can be behavioral. Behavioral changes are just as profound in terms of consequences for survival and population growth. Animal behaviorists were fascinated to observe that the Israeli black rat suddenly learned how to strip the seeds of Jerusalem pines by moving along the conical whorls in which these grow.[36] (A brute-force attack consumes more energy than the seeds yield in nourishment.) Once the animals made this discovery, they doubled the area that they occupied in a very short time. Researchers showed—by means of cross-fostering experiments—that the dietary change was not genetically based. Instead young rats learn the pinecone-stripping technique by watching their mothers.

The Israeli rats illustrate that behavioral adaptation has profound consequences for success of a species. Yet many biologists play down this connection because they see the world through the filter of genetic adaptation—a simplistic idea that has admittedly inspired much research in the field. However, gene selection gives only a fragmentary picture of how animals fit in with their environment, and it is relatively powerless to explain adaptive variation in human societies.

Why is it that one population eschews sex before marriage (as is common among Arabs in the Middle East) and another one promotes sexual experimentation by teenagers? Among the Gond tribe of India, for instance, male and female adolescents associate in a community house, known as a ghotul, where they are encouraged to enjoy free sexual experimentation without any guilt or recrimination.[37] Why adolescent sexuality runs to such extremes is currently unknown, but no evolutionary scientist can be happy with the one-word pseudoexplanation so commonly ban-died about by social scientists, namely "culture."

Gene selection does not work well for complex behavior, and the ingenious formulations of evolutionary psychologists, with their Swiss army knives, modules, and Darwinian programs, only serve to highlight the failures of genetic determinism. As already pointed out, there is no convincing biological evidence for any of these ideas. Yet there is no reason to throw in the towel.

When it comes to behavior, natural selection can act directly on the phenotype: there is no need for any intermediary selective system—whether genes or memes or anything else (memes being hypothetical units by which ideas are transmitted as an analog to natural selection via genes).[38] Individuals who acquire a useful technique for getting food, such as black rats learning to strip pinecone seeds in a conical whorl, leave more offspring than those who fail to master this technique. This method of feeding becomes more common due to the success of those practicing it. The behavior itself pervades the population in subsequent generations, being transmitted by social learning, particularly from moth-ers to offspring. Even though it does not have a genetic basis, it has nevertheless evolved.

Such behaviorally mediated evolution is by no means rare, although we know relatively little about it because genetically minded researchers had a different focus. Of the various examples that have been studied, some involve unusual feeding practices of specific animal populations. Chimpanzees fish for termites by dipping an object into an opening of the termite hill and waiting for the termites to grab on. Some chimps strip the bark of a twig before using it; others discard the stick and use the bark as the fishing implement. Which method is used may not matter very much except that the persistence of different techniques in local populations shows that these practices are socially learned—usually from the mother. Solving a problem of adaptation to the environment behaviorally can mean that a population either thrives or fails.

The point is illustrated by an endangered species: the Mauritius kes-trel. This bird saved itself from extinction by adopting the novel practice of nesting on cliffs. The kestrel faced extinction because its nests—origi-nally built in tree cavities—proved vulnerable to monkeys that had been introduced to the island by humans.[39] In 1974, one pair of kestrels nested on a cliff face that was safe from monkeys.

The young prospered and bred at the same site. This was possibly because they had developed a strong nesting preference for the place where they were raised. This sort of attachment to features of the early environment is called imprinting, and it is the mechanism through which young goslings follow the mother goose. Imprinting is an individual rather than a social form of learning. Yet the cliff-nesting phenomenon cannot be satisfactorily explained as happening in one generation and is therefore best thought of as social learning (i.e., a behavioral tradition). The cliff-nesting tradition saved this endangered species and was thus favored by natural selection without any change in gene frequencies. Of course, genetic evolution is not entirely irrelevant because the Mauritius kestrel evidently has an evolved flexibility in its nesting habits that might be affected by genotype, although this has never been demonstrated.

Like the Mauritius kestrel, humans possess an evolved propensity for a flexible response to environmental factors. This propensity generally increases our chances of survival and reproduction. Admittedly, modern fertility is often below replacement levels even after adjustment for low infant and child mortality.[40] Below-replacement fertility is a highly predictable feature of urban life in developed countries. The modern urban environment reduces fertility due to lack of affordable living space and other factors that make having children expensive, whereas the Agricultural Revolution boosted fertility due to increased food production.[41]

Such complex responses to environmental variation are predictable. Every agricultural society of the past had high fertility, whereas every developed country today has low fertility (with the exception of immigrants who are not yet fully adjusted to their new home and religious minorities who deliberately live an antiquated lifestyle, such as the Amish in Pennsylvania or the Hassidim in Israel who keep women in the home). Increased nutrition mostly improved health and reproductive success throughout human history, yet it undermines contemporary health due to overnutrition, reduced physical activity, and obesity-related illnesses in developed countries.[42]

Such responses to environmental variation are predictable, but these patterns are not attributable to gene selection. Instead, they are a feature of our behavioral adjustment to modern conditions. They are instances of adaptive variation rather than the genetic adaptation that helps match an animal's body shape, or internal physiology, to the way that it makes its living. Of course, adaptations themselves, considered as phenotypes, are never either completely environmental or completely genetic. How some phenotype emerges is very much determined by a range of environmental factors. Gluten sensitivity is a genetic disease that develops differently in different individuals from the extreme of celiac disease to milder variants. Of course, withdrawing gluten from the diet prevents symptoms from being expressed, and the same is true of other metabolic illnesses, such as phenylketonuria, which involves a serious toxic reaction to phenylalanine in the diet. How a genetic disease is expressed varies with the environment.

Adjustment to the modern environment is generally not gene based. Yet it is a patterned and predictable change in response to changing environments and thus bears all the hallmarks of adaptation, albeit by mechanisms that are just beginning to be understood.

ADAPTATION TO MODERN LIFE

Agriculture produced profound changes in human biology and behavior, including genetic adaptations to new dietary regimens, disease landscapes, and so forth. Humans were also profoundly changed by economic development in terms of health, stature, intelligence, psychology, and behavior. These adaptations to modern life were partly genetic but mostly independent of gene selection, considering the relatively small amount of time since agriculture became a key form of subsistence. The same is even more obviously true of the Industrial Revolution and the marked transition of societies toward excess production (and consumption).

As to the underlying mechanisms of these important changes in subsistence economies, evolutionary psychology is largely silent based on the presumption that humans are maladapted to modern life. I want to make a detailed case that humans, like all other species, are capable of exquisite adaptation to their *current* environments. Influential mismatch theories draw a different conclusion. Consequently, many evolutionary psychologists have ceded explanation of modern social behavior to cultural determinism.

Cultural determinism focuses almost exclusively on social learning. There is nothing wrong with social learning as such, but cultural determinists make a number of assumptions that stick in the gullets of evolutionary scientists. While I would prefer to avoid polemical statements, it is important to be very clear about what I see as problematic about cultural determinism as currently practiced by mainstream behavioral scientists. In addition, there is a small circle of evolutionary-minded cultural determinists who promulgate notions of human uniqueness and human superiority that animate titles such as *The Secret of Our Success: How Culture Is Driving Human Evolution, Domesticating Our Species, and Making Us Smarter.*[43] The author, Joseph Henrich, is a lot more scientifically sophisticated than most cultural determinists, but his ideas on human exceptionalism do not fit well with a natural-science approach to human evolution (see next chapter).

For the moment, I want to address some of the conceptual problems associated with the cultural determinism promulgated by mainstream, or non-evolutionary-minded, cultural determinists. One of the main conceptual problems with cultural determinism is the assumption that people are something entirely new or that we became different than the rest of the evolved world when we exerted more control over our environment and began communicating with grammatical language. In this new, exceptional state, human behavior and psychology are no longer shaped by the natural environment and are determined instead by what is being communicated between individuals and across generations. As Emile Durkheim, the early sociologist, expressed it, social phenomena require social explanations.

The problem with this formulation is that it is just too glib, leaving unaddressed the critical scientific problems of how humans first departed from the natural world and why. Answers to these questions seem possible only in the context of the same evolutionary science that describes all other species. Of course, humans are not alone in shaping their environment using tools or in having a complex communication system that transmits information between individuals and across generations. Chimpanzees make tools to catch termites deep in their mounds, for example. Both birds and cetaceans use complex communication systems that transmit individual and group information, as illustrated by songs and song dialects of birds. Ignoring these issues seems intellectually lazy. Moreover, evolutionary psychologists have always maintained that human beings make sense only in the context of evolution. Although that approach had seemed ineffectual while evolution was reduced to genetic determinism, inclusion of many new mechanisms for evolutionary change hold out the prospect of reincluding our species in the natural world. According to contemporary biology, genes are just one ingredient in the process of constructing phenotypes that are matched with an animal's way of life.[44] Of course, these also include social learning, which is believed to affect most, or all, vertebrate species.[45]

So the view that humans are qualitatively different from other species on the planet and that human life takes a completely new step into the realm of "culture" is, I believe, a fallacy. More important perhaps, it is a journey into mystical, scientifically hazy territory where easy circular "explanations" are preferred over a thoughtful analysis of the many mechanisms through which behavioral phenotypes are built. It is only when that is done that we can begin to entertain the possibility of humans doing something profoundly different from what all species have done since the dawn of life. Otherwise, accepting the circular reasoning that goes for cultural explanations commits the disciplines of human behavior to continue going around in explanatory circles without making any progress.

Just as agriculture altered the adaptive landscape humans encountered and produced a myriad of adaptive changes in biology, behavior, and even personality, it emerges that humans are profoundly altered by adaptation to modern economies. If economic development has many profound consequences for nutrition and fertility, it also transforms sexual behavior from patterns observed in agricultural societies. One notable change is the increase in sexual behavior of single people compared to a more restrictive sexuality for our agricultural forebears. The trend is most obvious in terms of premarital sexuality, which increased from a small minority of women a century ago to a substantial majority today. These issues are taken up later in more detail.

The purpose here is merely to show that human beings are constantly changing to meet new challenges in modern environments. As a result, human populations of developed countries are predictably different from farmers that preceded them. Just as they are taller, heavier, and more intelligent, they are also more sexually liberated and more geared toward gender equality and depart from sharply differentiated gender specializations. Such predictable changes in behavior in response to an altered social landscape satisfies the criteria for adaptive change. If so, then the key question to ask is whether such changes spring from novel mechanisms applying only to humans, or whether they reflect the same mechanisms through which other species become more closely matched to their way of life.

All vertebrate species show potential for both conflict and complementarity between the sexes as they mate and raise young. From this perspective, the shift to gender equality in modern societies may reflect declining gender specialization at work. This phenomenon also alters relations between the sexes. In particular, economically independent women do not have to compete so intensely for masculine economic investment and are consequently freer to express their sexual impulses. There is a corresponding change in premarital sexuality because market forces dictate a different equilibrium between men and women such that a masculine preference for sexual expression earlier in a relationship gets expressed.

Of course, similar market principles play out for other vertebrates where the reproductive system is affected by the supply and demand for males versus females. This generic explanation for shifts in gender behavior offers a viable scientific alternative to the reigning cultural determinist account (known as modernization theory, based upon changing values).[46] I argue that such cultural determinist accounts of social change are flawed and circular.

Studying sexual behavior as a market-driven phenomenon replaces top-down cultural determinist interpretations with accounts in terms of individual psychology. It embraces methodological individualism, or the idea that group trends reflect forces acting upon individuals. One advan-

tage of doing this is that the explanations so gleaned are clearly adaptational and can often be applied to many different species. For example,
among birds, the level of parental investment is predictable from the
supply and demand of males relative to females, as is true of our own
species. Similar explanations apply to the level of polygyny in the reproductive system.

The complexity of human adaptation to modern environments is greatly increased by the fact that we alter our own environments, and so
change ourselves. The domestication of animals posed new challenges to
our immune systems and led to partial immunity against some animal
diseases like cow pox. Nothing changed our species like the development
of a global economy that produced very rapid adaptations of our brains
and bodies by radically altering our way of life. Such phenomena are
generally excluded from biology because they fall squarely inside the
disciplinary boundaries of economics. Even so, they are relevant to our
profound adaptation to modern conditions, which includes increased
body size and longevity along with greater cognitive processing demands
and capacities. It is particularly important to include them in an evolutionary approach that is dynamic (in contrast to the static view of human
nature previously encouraged by evolutionary psychologists).

Humans are not alone in being very good at adjusting to ecological
variety. Many other species are opportunistic and colonize diverse habitats with obvious examples including some of our commensals, such as
rats, dogs, and cockroaches. A less obvious analogy is the locust that
changes its biology and behavior quickly and profoundly in response to
varying population density. Human adaptation to modern urban life has
many clear parallels with the transition of solitary grasshoppers to social
locusts, not least of which is an exponential growth in population. This
analogy captures the extreme change and instability characteristic of
modern life much better than social insects, like bees or ants, to which
human societies were often compared since the days of Aristotle and
before. Nothing characterizes recent human adaptation more than our
restless colonization of new lands and the subsequent development of
shipping, trade, and the globalized economy that offers a very different
subsistence economy and social landscape than foraging or farming.

The fact that economic growth has continued relentlessly ever since
the Industrial Revolution is not unrelated to human evolution as scholars
often assume. Indeed, a prosperous global economy greatly changes the
ecology of life on earth. This is becoming increasingly clear in relation to
damaging effects on the environment. Economic growth increases com-
bustion of fossil fuels, raises the amount of greenhouse gases in the
atmosphere, and increases average temperature of the atmosphere and the

seas, issues that are becoming of increasing concern due to extreme weather events and sea level rise.

Of course, these changes are not just challenging to human life but threaten entire ecosystems, such as rain forests and coral reefs, with mass extinctions unlike anything seen since the human-induced loss of large animals to overhunting in the Pleistocene Overkill some forty thousand years ago. Interesting and important as these topics are, their human impact is likely to be much more important in the future even than now. They are outside the scope of this book, which is much more concerned with how global growth has affected the adaptive landscape up to this point.

Economic growth changes human biology and behavior in profound ways. One marked difference is the increase in life expectancy, which almost doubled in the United States during the twentieth century. Improved nutrition also led to substantial increases in stature of around 10 percent, a pace of skeletal change that is unprecedented in human evolutionary history. Thanks to a more sedentary lifestyle and increased intake of high-energy foods, modern populations have become increasingly heavier until overweight or obesity characterize the majority of the population, something that is of great concern for health reasons but is also without historical parallel. There is little or no overweight among indigenous people who lead very physically active lives.

These changes are substantial enough that a hypothetical biologist from another galaxy might categorize urban residents and members of subsistence societies as different subspecies. Beyond basic anatomy and physiology, a psychologist from another galaxy would observe huge differences in behavior that go beyond patterns of locomotion and rest. Indigenous people are motivated by the same goals as other primates: they struggle to find food, to produce offspring, and to raise them to maturity. In contrast, many members of developed societies do no food procurement, and some even raise no offspring by choice. Whereas members of subsistence societies have limited goals and are generally more content with the status quo, people in developed societies strive for higher social status in a steeply hierarchical, wealth-based social system and wish to impress others with their wealth or achievements in sports and performance arts, academic life, medicine, technological innovation, government and politics, or the creative arts.

As societies develop, members become increasingly removed from basic biological motives and much more preoccupied with new, more abstract concerns having to do with making an impact in some specialized field of endeavor. At the same time, we become much more tolerant of diversity, a disposition that is increased by prosperity and tuned down by economic recessions. According to modernization theory, such changes may be attributed to relevant changes in attitudes that are assumed to be a product of cultural diffusion (through channels that are rarely specified or quantified).

I make the case that modernization theory fails to pass muster as a scientific explanation, principally by being circular and failing to separate causes from effects. In its place, I propose an adaptational theory of social change that is founded on well-established principles of animal behavior. For instance, mammals who are raised in secure situations with reduced predator threats are bolder and more adventurous and develop superior learning capacity. This basic mechanism is invoked when humans experience more favorable economic conditions that leave them more open to intellectual exploration and more tolerant of diverse perspectives.

Resistance to adaptationist perspectives on human societies has many sources, but one of the biggest obstacles is probably historical. A straight line is drawn from Herbert Spencer's discredited theory of Social Darwinism (summarized in the phrase "survival of the fittest") to the unfortunate consequences of Nazi "race science," including the Holocaust. Such theories were certainly floated by Darwin in *The Descent of Man*, but they are now deservedly deposited on the trash heap of history.[47] Another major obstacle is a strong, seemingly unshakable belief in human exceptionalism. We are different from, and superior to, all other species! We have culture! We have language! What is the objective scientific evidence that humans are qualitatively different from other mammalian species thereby justifying the exclusion from natural-science explanations of our extraordinary historical trajectory from foraging to trading? These issues are taken up in chapter 4. Before broaching such issues, it is important to ask the question of how individual differences, and hence species differences, arise in the lifetime of the individual.

3

EVOLUTION AS A
DEVELOPMENTAL PROCESS

Darwin's theory of evolution is an attempt to explain characteristics, such as the long necks of giraffes. These are *phenotypes* as opposed to genotypes. He really knew little about the underlying biological mechanisms, and developmental biologists are still coming to terms with their Byzantine complexity. Ironically, modern Darwinism, including evolutionary psychology, is preoccupied with genetic determinism thanks to a paradigm shift in biology in the 1930s known as the New Synthesis, which integrated field biology with population genetics and set out to explain changing phenotypes in terms of changing gene frequencies, thereby reducing evolution by natural selection to genetic determinism. The logic is powerful because genes are, after all, the primary mechanism of trait inheritance and trait inheritance is the cornerstone of Darwinian evolution. So, from a theoretical perspective, modern genetics combines the powerful mechanism of natural selection with the equally potent force of genetic inheritance.

The New Synthesis was powerful enough to invade the social sciences in the sociobiology revolution that is summarized in Edward O. Wilson's text of that name, which was made notorious by its final chapter dealing with humans.[1] Weaknesses in sociobiology were immediately apparent to critics, however. Social scientists batted away the new approach by linking it to Nazi race science and the eugenics movement founded by Darwin's cousin Francis Galton. Wilson's case was greatly weakened by the fact that he knew relatively little about human psychology, being an entomologist who specialized in the social insects, and had limited expertise in the molecular genetics being used to explain complex human behavior. However, it is sometimes true that fortune favors the bold, and

Wilson's claims earned him the most fame and notoriety for any field biologist since Darwin himself.

Just as Darwin was not above filling in conspicuous theoretical gaps with imaginative fiction, Wilson popularized a very loose way of thinking about evolution. For example, he made questionable claims about genetic inheritance—such as the view that there is a gene for morality—made in the absence of any concrete evidence about which genes are implicated or how they act and at what point in development and without a meaningful definition of moral behavior, something over which no two philosophers can agree.

Genetic determinism is favored by its dramatic simplicity. Asked why a giraffe has a long neck, a genetic determinist can say that the crane-like structure supporting the head is an invariant feature of giraffe genetics. This certainly has some truth to it but unfortunately omits all of the hard scientific questions. Which genes are involved? How are they activated? When are they activated? What is their biochemical impact? How does this affect neck length? How does giraffe neck development differ from neck development of related species? How is neck length affected by nutrition? How do giraffe hearts manage to pump blood to the lofty head? Without addressing all of these problems in some way, the bald assertion of genetic influence means little. At most, it points in the direction of a large number of scientific problems needing attention.

Simple genetic determinism works well in mathematical models but has never been credible in concrete examples where the welter of biochemical effects on some trait in the course of development is apparent. The same is even more true of behavior where environmental inputs through the senses are always so important and where the phenotypic behavior seems quite distant from protein synthesis. The fundamental problem with simple genetic determinism is that it fails to grasp the true importance of developmental mechanisms. After all, a gene is visible to natural selection (so to speak) only if it has some impact on the end product of a complicated developmental unfolding.[2]

Sociobiology always lacked scientific plausibility simply because it explained so little. No one ever believed that there was a specific gene for morality or aggression. If not, then using that language has little foundation in reality, for a skeptic can say, "Identify the gene! Show me exactly what it does!" and so on. Evolutionary psychology is a more sophisticated version of sociobiology that abandons simple gene effects on behavior and asserts that genetic influences on psychology are baked into the brain early in development.

The central idea is the Swiss-army-knife model of brain function. Leda Cosmides and John Tooby claimed that the brain is composed of a number of special-purpose programs that were designed by two million years of human evolution to solve recurrent problems faced by our forager ancestors.[3] So there was a module for detecting cheats, another for preventing sexual infidelity (i.e., jealousy), a third for investing selectively in close relatives, a fourth for finding the best mates, and so forth. Such *Darwinian programs* may be expressed differently by gender (with men being more sexually jealous than women, for instance), but they emerge similarly in all societies, resulting in human commonalities across societies.

This theory is attractive in its simplicity, but it is falsified by developmental evidence. The assumption that genes can program behavioral phenotypes is simply a misunderstanding of how brains develop. It is true that the brain does have specialized local functions, as illustrated by sensory maps in the parietal cortex, but these do not follow any genetic blueprint. It has become increasingly clear that the brain develops its specialized function thanks to experience. For instance, when a person learns to read, their capacity to recognize words improves thanks to specialization in an area of the brain that is not specialized in nonreaders.[4] This "letterbox" develops throughout childhood and into adolescence as the person's reading facility improves.

So the brain clearly has refined functional specializations, but these are not—and cannot be—programmed by genes according to any mechanism studied by developmental neuroscientists.[5] The idea that they are the product of gene selection is thus clearly false. Obvious though that conclusion might be today, it was far from clear when the theory was proposed several decades ago. Moreover, it seemed to derive firm support from two branches of psychology, namely cross-cultural psychology and behavioral genetics.

Cross-cultural comparisons, based on data collected in thirty-seven countries, found a pattern of gender differences in mate selection criteria that were consistent with predictions from evolutionary psychology.[6] In particular, men emphasized beauty of a partner more strongly than women, whereas women valued economic success in a future husband more than men did in a potential bride. Such phenomena were assumed to be rooted in biology and thus possibly open to some form of genetic influence. Of course, genetics cannot have any kind of simple effect on gender differences in mate selection criteria. The reason is that men have almost exactly the same DNA as women, the major exception relating to the Y-chromosome that determines biological sex. It is possible, however, that sex hormones or neurotransmitter actions influence these behaviors. If so, then they might plausibly be affected by receptor populations that are themselves affected by genetics. Although implausible, this is at least theoretically possible.

The second problem with cross-cultural universals in mate selection is even more troubling. That is that the supposed universals are vanishing before our eyes. As women become more active in paid employment, one finds that their sexual psychology is converging with that of men. Of course, that fact falsifies the entire thesis. Why do women become more interested in sports competition and more sexually liberated in modern societies? To an adaptationist open to nongenetic explanations, the answer is quite obvious. They behave more like men because their occupations and social impact are growing closer to those of masculine occupations and social positions. At the same time, there is a decline in gender specialization at work that becomes more broad-based as men assume greater responsibility for domestic work and caring for children.

If so, perhaps gender-related division of labor was always more of an immediate response to practical contingencies in the everyday lives of human populations. Ever since Adam Smith, we know that the division of labor can yield tremendous increases in efficiency.[7] Women were always more effective as mothers, whereas men made more efficient hunters of large game. In an ecology where large numbers of children must be produced to avoid population extinction, any community where men did the child care and women did the hunting would be at a competitive disadvantage. Similarly, once humans became involved in intercommunity warfare, those groups that sent their women to war would soon lose out because the number of women, rather than the number of men, determines the level of reproduction and population growth rate. The fact that women so rarely fight in wars shows adaptive design, but this is not because there is a gene for feminine pacifism. Women are clearly not pacifists. Indeed, if one looks to verbal aggression, as well as physical aggression in relationships, some scholars conclude that women may actually be quantifiably more aggressive than men.[8]

In a similar vein, it can be argued that the division of labor in agricultural work reflected seemingly permanent social realities. First, women needed to produce several possible heirs to maintain their farms due to high mortality rates. Second, men had greater upper-body strength, making them more suited to heavy agricultural work. Given these practical realities, it would not be efficient for men to specialize in child care while women tackled the heavy farm work.

There are exceptions, of course. Throughout sub-Saharan Africa, women do much of the agricultural work, whether it is tending garden plots or looking after domestic animals. Interestingly, African women are taller and stronger, on average, than their European counterparts who do less heavy agricultural work.[9] Men of the region do less work on farms than their European counterparts in primarily agricultural societies did in the past. This difference is not well understood but likely relates to an imbalance in the sex ratios. In Europe, there are much more men than women, with a birth sex ratio above 106 males per hundred females and comparatively low male mortality. In Africa, there are only about 103 male births per hundred females and males have higher mortality rates. As a result, there is a scarcity of adult men so that men control the mate market and are therefore able to negotiate marriage on their own terms. Instead of making a heavy economic investment in their wives and children, as European farmers did, they can select hardworking women who take care of a lot of agricultural work, in addition to raising children.

Another intriguing exception relates to Neolithic farmers in Europe. Recent archaeological finds indicate that women had highly developed upper-body musculature that is comparable to competitive female athletes of today.[10] The inference is made that they were accustomed to some heavy lifting on the farm, although we do not know why. It is possible that males had high mortality from warfare in this period given that there was conflict over the highly fertile loess soil in the region. Whatever the reason, gender work specializations are far from being fixed. Like everything else, they are subject to adaptive variation in response to environmental differences. If you expect universals, expect to be surprised! The story of universals may now be weakened by the reality of dramatic changes in the division of labor by gender, but evolutionary psychology still derives strength from an outsized faith in genetic determination of behavior. As pointed out, this approach is coming under withering criticism from developmental systems theorists. Genetic determinism draws its strongest support from modern behavior genetics that analyzes personality traits and abnormal psychology from the perspective of family resemblances.

BEHAVIOR GENETICS

We all know that children resemble parents in intelligence, personality, and emotionality in addition to being physically similar in traits such as eye color, stature, and body build. The resemblance between children and parents cannot be attributed solely to genetic inheritance. After all, children consume the same diet as their parents, live in the same home, and are socialized by their parents in similar ways to how the parents themselves were socialized. Some traits are familial without being entirely genetic. For instance, intelligent parents provide a stimulating environment for their children simply by talking to them.[11] One way of teasing apart genes and environment is to study populations varying in degrees of relatedness that either grow up in the same environment or in different environments. Of course, it is impossible to assign humans randomly to rearing conditions, but the next best thing occurs when infants are adopted out soon after birth.

The Minnesota Twin Study, looking at identical and fraternal twins reared apart, offered compelling evidence that offspring resemble parents due to their genetic composition. [12] In general, identical twins reared apart were much more similar in personality compared to fraternal twins reared apart, suggesting that shared genes play an important role in intelligence, sociability, emotionality, extroversion, and so forth. One way of assessing the role of genes is to compare the correlations between identical twins that share 100 percent of their genes and fraternal twins that share just 50 percent of their genes. (Heritability may be calculated as twice the difference between correlations for identicals and fraternals.) For many individual differences of interest to psychologists, heritability estimates hovered around 50 percent, suggesting that genetic influences are approximately equal to environmental ones.

Unfortunately, subsequent research has not nailed down the biological mechanisms through which genotype affects brain biology and personality. Specific neural receptors are affected by gene expression, but these effects account for a relatively small fraction of genetic variance in personality. Another problem is that, as a practical matter, it is surprisingly difficult to differentiate genetic and environmental influences on personality.

One serious complication is the fact that identical twins reared apart gravitate to similar environments that magnify their early temperamental similarities. If they are relatively fearless as toddlers, they end up in riskier sports and occupations that increase their risk taking, for example. If they are highly intelligent, they spend more time reading, which increases their intellectual capacity. What had looked like straight gene effects are really due to an interaction between genes and environment. Of course, modern developmental theorists would argue that separating genes and environment is a quixotic endeavor because both influences are at play in the unfolding of every trait, whether behavioral or somatic.

This conclusion seems obvious today, but it was less obvious in the 1970s when behavior genetic evidence provided comfort to evolutionary psychology by suggesting a large role for genes in determining personality and intelligence. The Minnesota Twin Study had suggested that identical twins reared apart were spookily similar in their interests and behavior. This impression was strengthened when photographs of twins that had been separated from birth showed that they looked strikingly similar and even wore similar clothing. Of course, some identical twins are not much more similar in appearance than same-sex fraternals, but they did not make it to the newspapers. Unfortunately, a picture is worth a thousand words, and the journalistic reports were used to convey the impression that personality is fixed at birth by genes. The pictures were backed up by numerous anecdotes such as one about a pair of twins being the only ones to build a circular seat around a tree in their back yard.

Entertaining as such stories are, no behavior geneticist ever seriously believed that there is a genetic basis for building circular seats around trees. All such anecdotes are compromised by (a) failure to consider negative instances and (b) inability to assess the base rate for the behavior of interest in the relevant population. Then there is the fine detail of which genes might be responsible, when they are activated in development, and how.

Behavior genetics research indicated that approximately one half of the individual differences in personality are attributable to genotype. Of course, some traits, such as creativity and religiosity, have much lower heritabilities (30 percent and 0 percent respectively). [13] The strong influence of genotype on personality traits was hardly surprising. After all, temperament was of long-standing interest to breeders of horses and dogs who were selectively bred for activity level, docility, and trainability. Personality tests are designed to tap into such stable differences in humans that might also be expected to have a heritable basis if only because temperament is heritable in other species.

Behavior genetics research suggested that environmental influences on personality traits are roughly equivalent to genetic influences in terms of population differences explained. That is a salient conclusion in its own right because it indicates that there is always some malleability in psychological development. That is what *should* happen if, as this book argues, we are continually experiencing adaptive modification that helps us to match our current environment. This impression is also supported by the precise nature of environmental influences on personality development.

The impact of the environment on development is divided into two types. Either we are affected by the environment that we share with siblings growing up in the same household, or we are changed by the unique experiences we have in that setting. Contrary to the assumptions of many social scientists, behavior geneticists reported that what really mattered for personality development was the unique environment we experience as individuals, rather than the one we share with our brothers and sisters. Indeed, they concluded that the shared environment is relatively trivial, accounting for a mere twentieth of individual differences for most traits studied.

Two qualifications are in order. The first is that most behavior genetic studies are restricted to middle-class families in the same ethnic group because they rely on the natural experiment of adoption. This means that the home environments are generally good enough to favor academic success and do not tap into the extreme poverty and deprivation experienced by some underprivileged groups.[14] The relevant empirical evidence suggests that children from underprivileged homes who are adopted into more favorable economic conditions do indeed perform substantially better on IQ tests, although we need to be cautious about drawing such conclusions because the biological impact of the shared environment can take multiple generations to play out based on effects of famines, for instance.

Adoption studies mostly focus on middle-class families of European descent who are selected to provide an adequate home environment for the intellectual development and general health of their children. They exclude potentially abusive parents as unsuitable for adoption, for instance. This means that the sort of environment that generates extreme stress leading to low intelligence, ill health, psychopathology, and crime is systematically excluded.

If we wanted to criticize the early research in behavior genetics that found no substantial effect of the shared environment, we could point out that environmental differences were generally not substantial enough to be detectable. This would be analogous to dismissing the effectiveness of a drug because we had only tested it using a single tiny dose. Moreover, there is compelling evidence from other lines of research showing that children growing up in extremely stressful homes score substantially lower on IQ tests, have increased risk of psychopathology and crime, and suffer more from chronic health problems that reduce their life expectancy.[15] No reasonable person would dismiss such effects as trivial, but a lot of twin and family studies are blind to them.

Another major complication is the influence of environmental variation on gene expression (or epigenetics) that has waxed in the past two decades even as interest in simple genetic determinism of behavior has waned. How genes are expressed in future generations is affected by current experiences with important behavioral consequences. This occurs when genes are marked in ways that are transmitted to offspring.[16]

One result is that animals exposed to frightening environments, such as those having high predator density, grow up to be more fearful and less willing to explore novel environments.[17] This can be interpreted in adaptive terms given that more cautious individuals are more likely to escape predators and survive to reproduce. Be that as it may, it is hard to imagine how this mechanism could have been genetically selected given that it occurs too rapidly for evolution by gene selection to occur.

However epigenetic adaptations come into being, there is plenty of evidence that they exist and are a factor in cross-generational similarity. Although direct experimental evidence is ethically impossible, some human experiences are analogous to animal experiments. For instance, children exposed to traumatic experiences in childhood are less willing to explore their environments, judging from the fact that they score lower on IQ tests.[18] This has profound implications for child-rearing practices, crime, and psychopathology.

CHILD ABUSE AND THE EPIGENETICS OF VIOLENT CRIME

Children who are abused may get locked into a cross-generational cycle of violence. Indeed, the majority of abusive parents were abused themselves as children.[19] The underlying biological mechanisms likely include epigenetic change, in other words, acquired alterations in genes that can be transmitted across generations.

Some insight into these mechanisms was provided by experiments with rats. This research began with the observation that some rat mothers are more attentive to their offspring than others are. The amount of maternal care female pups receive affects their attentiveness to their own offspring once they become mothers. It is important to realize that such variation is likely an adaptive response to early environmental conditions affecting the mother herself. It makes no sense to cast moral aspersions on nonhuman mothers, and the same approach is applicable to humans.

The "abusive" tag is thus unfortunate, but we are forced to follow common usage, particularly given that it is the only descriptor that is widely used and readily understood. "Abusive" describes reduced care and emotional closeness, combined with increased aggression and rejection. For rats, maternal attentiveness and care are measured in terms of the time spent licking pups, which varies not just between mothers but also between individual offspring, with males receiving more care than females.

Maternal licking affects the pattern of DNA methylation in pups' brains so that there is greater expression of glucocorticoid receptors.[20] Behaviorally, greater expression of glucocorticoid receptors means a greater capacity to deal with stressful situations while remaining calm. So rats receiving a lot of maternal licking are less fearful in novel situations. It is not hard to understand that this disposition is more suited to a safe environment with few threats from predators, for example, than it is to a risky environment having high mortality rates.

For humans, childhood abuse was associated with abnormal methylation in the adult brain according to analyses of suicide victims.[21] Specifically, there was greater methylation of the glucorticoid promoter in the brains of suicides who had experienced childhood abuse compared to those who had not. As a result, they had *decreased* glucorticoid receptor expression (given that methylation of genes silences them). Research on rhesus monkeys found that abusive early experiences alter the brain in similar ways to those recorded for rats and humans.

Just as some rats make more nurturant mothers than others—who spend more time licking offspring—some rhesus monkeys handle their infants more roughly than others. Monkeys exhibit a similar cross-generational pattern as observed for humans. Intergenerational transmission of infant abuse in rhesus monkeys is the result of early experience rather than simple genetic inheritance according to cross-fostering experiments.[22] So offspring of nurturant mothers that were raised by abusive females grew up to be abusive mothers. This is likely due to a combination of social learning and altered DNA methylation that is transmitted to future generations through genes.

In rat experiments, pups exposed to abusive maternal care had increased methylation of the BDNF gene in the frontal cortex. In humans, this methylation pattern is associated with major psychoses including schizophrenia and bipolar disorder. So abusive maternal care has substantive effects on the developing brain that make individuals more vulnerable to stressors later in life. One implication for humans is that there would be reduced impulse control and therefore greater likelihood of committing serious crimes, including crimes of violence.

Evidently, maternal behavior is epigenetically calibrated to match prevailing social conditions, much as metabolic systems are affected by the food supply before and during gestation in ways that make adaptive sense.[23] Human maternal behavior likely responds to psychological stress due to danger, hunger, or competitive aggression. Insensitive parenting is correlated with offspring delinquency and crime.[24]

Children of psychologically abusive parents are likely to be sexually active from an earlier age. They are also liable to be more impulsive in their sexual behavior, exhibiting a pattern of short-term or low-investing sexuality that is partly explainable in terms of the epigenetic impact of stress hormones on the developing brain.[25] The adaptive rationale for this pattern is that difficult and stressful environments favor earlier reproduction. Of course, impulsive sexual behavior is correlated with increased risk of criminal offending, particularly because these go together in defining teenage delinquency.

Harsh parental practices, such as corporal punishment, are transmitted across generations and are surprisingly resistant to change. Even after they are alerted to the adverse effects of coercion and corporal punishment for children, parents in disadvantaged communities continue to use those tactics, in part, perhaps, because that reflects how their brains were affected by childhood experiences. Detailed training in more empathic parenting produced no change in parental practices.[26]

It seems that the more stressful conditions of low-income neighborhoods elicit more corporal punishment and less empathy compared to a more comfortable middle-income neighborhood and that these predilections persist throughout adult life, reflecting brain neuroreceptor populations and neurochemical functioning.[27] The outcomes of these varied parental practices may shape adults who are better suited to their specific social niche. Children of abusive parents have more trouble controlling hostile impulses and are at increased risk for criminal offending and impulsive behavior more generally.[28]

So, even within the realm of genetic influences, one finds a much faster calibration of behavior to varied environmental conditions than we might expect from studying anatomical adaptations that stretch out over tens of millions of years. Of course, we still do not understand the evolution of epigenetic mechanisms themselves. One can get around this difficulty by focusing on the match between phenotype and function, which is precisely how all adaptations are defined, from the neck of the giraffe to the song of the lark. Lark song is partly learned, of course, and the song is acquired from local males using local dialects. This phenomenon highlights the complexity of behavioral adaptations: they draw on many distinct mechanisms apart from genetic inheritance. However the phenotype is built does not really matter; it is still subject to natural selection.

In an astonishing recent confirmation of previously dismissed Lamarckian ideas, the experiences of parents may indeed alter the biology of offspring. This adds another dimension to the tortuous problem of why children resemble their parents. We see that far from being separated into discrete components, genetic and environmental influences on behavior development are intermingled in ways that challenge the validity of earlier confident conclusions in behavior genetics on the neat partitioning of genetic and environmental influences on some trait of interest. In reality, genes and environment both affect the development of every trait because environments affect the biochemical mix in which genes are activated or silenced.

This drama is enacted according to a relatively predictable series of events, however. What transpires currently is affected by what had happened earlier. This sequence is not a genetic program, however, so much as a natural unfolding of the action. Hamlet cannot die in the opening scene because he has a number of depressive soliloquies to deliver first, and he cannot survive the last because it is, after all, a tragedy. This fundamental plot logic also applies to biological development whether of body or behavior and is illustrated here by body shape (morphology) and by gender.

MORPHOLOGY

That evolution functions at a developmental level is brought home by the study of bodily development. The basic reason is very clear. Gene-based evolution consists of descent with modification. That modification must somehow be accomplished during development from the fertilized cell to the adult form (or ontogeny).

All vertebrates are descended from fish. This means that the early development of a mammal or bird is distinctively fish like. This is not because the process of development recapitulates the details of vertebrate evolution. Ontogeny does not recapitulate phylogeny with the fidelity of a fine sound recording. Indeed, there is no such record or program, and there could not be because the process is too complex to fit within the genome or any other biological information store. [29]

Beginning with a fish ancestor, how does one build on all of the features of a reptile, mammal, or bird? The solution is quite evolutionarily conservative. Ontogeny begins with the ancestor embryo and slowly constructs the newly evolved forms. For example, the human embryo begins with gill-like structures designed to take oxygen from water and modifies them into air-breathing lungs that function in much the same way given that they remain moist in the course of oxygen extraction from the air.

Beginning with the ancestor and using ancestral forms to mold current species is the basic process in development, and in that limited sense, it mirrors evolutionary history. This phenomenon leads to some very odd, seemingly inefficient phenomena, but we can be assured that it works best this way or something better would surely have replaced it. So the idea of beginning with the plan for a modern species and constructing it from scratch is a nonstarter. Just as mammals begin with fish-like forms, marine mammals like whales preserve many distinctive mammalian features, such as breathing air in lungs. Breathing in this way is highly inefficient for marine mammals that must ascend from the ocean depths every few minutes to get air. The fact that whales do not evolve gills that are much more suited to marine life reflects their ancestry but is curious and offers an intriguing clue about how development typically works.

The problem is that early mammalian development changes gill structures into lungs. Once the lungs are differentiated, it is difficult—or impossible—to reverse the process. Instead of moving as the crow flies, so to speak, the developmental process is analogous to a traveler meandering through unknown roads without a map. Where they end up is determined by which intersection they take at every point along the journey.

In addition to being committed to breathing in air, whales have a number of additional distinctly non-fish-like features that define them as mammals. They are warm-blooded. They even have a few vestigial body hairs. Moreover, they actually have four limbs, although these are modified into flippers. Why this is the case has become clear as their phylogenetic history was unraveled. Modern whales that are fully committed to life in the water had ancestors that were smaller and could move on land, possibly to escape large predators like sharks or perhaps even to breed after the fashion of many other marine mammals such as seals and sea lions. With increasing body size, they became immune to large predators, if not to small parasites, and had no need to visit the land so that their limbs could be fully committed to being flippers.

Geneticist Francois Jacob famously stated that evolution works as a tinkerer rather than an engineer. By this he meant that the process of adaptation involves altering materials already present rather than building from scratch using a blueprint. Of course, evolutionary tinkering is a developmental process because the phenotype is built during development. This principle explains some otherwise bizarre and puzzling features of anatomy.

The mammalian ear is one notable product of evolutionary tinkering. Our auditory system is a rather freakish phenomenon from the perspective of auditory engineering. It nevertheless works extremely well as illustrated by animals such as dogs whose sensitive hearing helped humans detect prey animals from a distance when they were domesticated some fifty thousand years ago. Humans do a great deal with their sound detection that facilitates a very refined language communication system, but canine hearing is probably some ten thousand times more sensitive in the technical sense that they detect much lower sound pressure levels. While humans are lucky to detect large animals moving a hundred yards away, dogs can hear them literally for miles. [30]

The ear is a fantastic Rube Goldberg device whose complexity and seeming redundancy can only be the product of a series of evolutionary accidents. Sound enters the auditory canal and vibrates the eardrum. This movement is transmitted through the inner ear by a series of three bones (maleus, incus, and stapes, or hammer, anvil, and stirrup) to the inner ear where the displacement produces a standing wave within the auditory fluid that bathes the hair cells. Movement of the hair cells produces an electrical stimulus that is transmitted, via the auditory nerve, through several way stations until it reaches the auditory cortex and produces sound perception. Imagine a human engineer that proposed developing an auditory detector according to this scheme! The story gets even more convoluted when one looks at the evolutionary history of some of these structures. Hair cells are presumably derived from sensory hairs on the external surface of the body. The bones of the middle ear are evolutionary leftovers. They used to function in the disarticulation of the jaws through which reptiles like egg snakes could swallow prey bigger than themselves. With strong chewing jaws, mammals could dispense with such niceties that became surplus material used by evolution the tinkerer to fashion the mammalian ear.

The improvisational nature of the evolutionary process is not just something that plays out over long time scales of genetic evolution. Development also proceeds according to a complex welter of contingencies. This phenomenon is manifested in limb development, which has some well-developed ground rules for a specific class of animal but can be altered by diseases and parasites.

Amphibians, such as newts, who lose a limb can develop a perfect replacement at any time of their lives, but mammals who lose a limb as adults are permanently disabled. If the amphibian is infected with parasites (nematodes), the replacement limb may be severely deformed, by having the wrong number of digits or even forming as two mirror-image structures.

Body structure is strongly affected by a small number of genes (Hox genes) that affect the positioning of limbs and ribs. Limbs and ribs have a curious antagonism in the sense that animals without limbs, such as snakes, have a large number of ribs and an elongated thorax to which the ribs are always attached. The limbs themselves develop from embryonic buds that can be moved around in experiments on embryos and thereby produce freaks with out-of-place appendages, such as a fly with a foot growing out of its head.

Once limb development begins, it proceeds in a predictable sequence progressing from proximal to distal. In leg development, for example, the femur develops first, followed by the tibia and fibulae, which are then followed by ankle bones, foot bones, and, finally, the toes. One curious feature of the development of limbs, and appendages more generally, is that the amount of growth is dependent upon local availability of resources. If two organs compete for bodily resources, one is liable to be smaller as a result. This phenomenon emerges in the case of horned beetles, some of which have impressively large horns that are used in competition among males.

Researchers discovered that moving buds early in development can have interesting results. Working at the University of Maine in the 1930s, Franklin Dove moved the horn buds of a calf from the normal position to a central location on the forehead.[31] As he anticipated, the two horns fused, producing an unusually large central horn that looked like the mythical unicorn. Whether unicorns ever occur in nature is unknown, but anomalous development and the resulting creatures that used to be displayed in circus shows are not all that rare. Although circus freaks look different in some respects, their development is completely normal in others, particularly in the functional adjustment of which they are capable. In that sense, they reveal a lot about how development achieves the mission of producing individuals who are well adapted to their environments.

FREAKS OF NATURE

One aspect of abnormal development is that people with what look like severe limitations may not be disabled at all. They are perfectly functional. People born without arms or legs make remarkable adjustments early in their lives so that they can maximize their capabilities and do not seem functionally disadvantaged at all. For example, a man born without arms or hands mastered the use of his feet to cut out valentines using scissors. [32] Of course, matters are very different for adults who lose limbs in an accident. They are not only severely incapacitated in the struggle to carry on a normal life but also suffer from persistent phantom pain in the missing limbs, something that does not happen for people who are born without limbs.

Of course, the same is true of other species, and this offers key insights into how movement develops. This phenomenon is well illustrated by four-legged animals, such as dogs and goats, in which the upper limbs are absent due to developmental anomalies. These freaks are capable of astonishing adjustments that permit them to walk functionally on their two good limbs. This is a remarkable accomplishment of the developmental process, given that bipedalism among humans is a slow evolutionary process that proceeds from knuckle walking by other apes, such as chimpanzees, to obligatory upright walking based on numerous changes in bone structures, muscles, ligaments, nerves, and central nervous system controls for movement. Yet the phenotypic "mutant" goat achieves this adjustment in the lifetime of the individual.

How is this accomplished? If it takes millions of years for humans to become bipedal as comparative anatomists conclude, how could a goat or a dog accomplish bipedalism within a single year in the absence of relevant gene selection of any kind? Mark Blumberg believes that this phenomenon bears testament to the remarkable capacity of abnormal individuals to respond functionally to changed conditions in early life, in this case the absence of upper limbs. He argues that in this respect they are no different from normal animals. This point is brought home by the development of locomotion in the jerboa, an animal that has large hind limbs and small forelimbs. The adult jerboa is effectively bipedal, moving through the air in large bounds similar to a kangaroo.

As the young jerboa develops, it starts to move in much the same fashion as other mammals. It begins by moving its front paws so as to rotate around its rump with inactive rear limbs in a pattern referred to as "punting," which is good enough for shifting position within the nest. As its limbs strengthen enough to lift the weight of the body, it acquires a conservative-mammal quadruped gait. As the hind legs grow, quadrupedal movement gets increasingly uncomfortable and the gait is adjusted. By day 10, the hind limbs are already much bigger than the size of the upper limbs, making quadrupedal motion very awkward because the longer hind legs tend to dig into the ground unless they are tucked closer to the body as the jerboa learns to do. Soon after day 40, when the hind limbs are already three times as long as the upper limbs, the animal no longer finds it possible to maintain quadrupedal walking and transitions to bipedal movement.

For jerboas, as for goats born without upper limbs, bipedalism emerges naturally as a response to the prevailing conditions for effective movement. Hence, Blumberg's argument that freaks, like the two-legged goat, are perfectly functional because they experience normal motor development, which is to say that they respond effectively to the developmental contingencies affecting movement. Of course, they do so due to feedback from their environment, rather than as a consequence of some shadowy motor program supposedly contained in the genotype and impressed upon the developing central nervous system via unknown mechanisms.

Movements may sometimes *seem* predetermined, as illustrated by the newborn step reflex described by child developmentalists (the plantar reflex). Although this may seem biologically determined, we need to be cautious. After all, the child has likely been kicking in the womb for several months and thus pushing the sole of its foot against resistance, which is precisely what the plantar reflex demonstrates. When the fixity of movement patterns is experimentally tested for other species, the concept takes a beating.

The issue of motor programs was investigated experimentally using rabbits. This species is of interest because the movement pattern generated in the spinal cord is different from that produced by the brain. The spine is characterized by typical walking movements of mammals where movement alternates between left and right limbs. On the other hand, the motor program controlled by the brain is simultaneous movement of left and right associated with characteristic hopping movements of the adult rabbit. Researchers wondered whether rabbit movement is autonomous, as biologists had often assumed, based on experiments that demonstrated movement patterns in the spines of dogs and other animals.[33] They severed the spinal cords of two-day-old rabbits and provided different types of movement stimulation. At ten days old, rabbits were strapped in a harness and their hind feet secured to pedals. The pedals delivered either the simultaneous movement characteristic of hopping or the alternating movement associated with walking. Fifteen-minute training sessions continued until the rabbits were thirty days old. In the test, rabbits were induced to kick their legs by pinching the tail. Would they produce the alternating movement pattern that was assumed to be preprogrammed in the spinal cord? The researchers found that the movement type was always the same as that with which they had been trained. They concluded that instead of being already present in the spine at birth, movement patterns of the rabbit are constructed due to early limb movements. Evidently, rabbits first develop typical mammalian walking patterns that are organized in the spinal cord—a structure that gets myelinated before the brain. Subsequent hopping actions get stored in the brain that is now sufficiently mature, and that brain pattern predominates in subsequent movements, likely because the brain characteristically inhibits lower control mechanisms in the nervous system.

The rabbit experiment agrees with observations on movement development in the jerboa. Both sources of evidence confirm the sensitivity of locomotor development to sensory inputs as these vary with bodily growth. Behavioral development is thus a complex, environmentally contingent phenomenon where the individual animal makes the best adjustment it can to the changing challenges of efficient movement around its environment.

Why development proceeds in this way seems obvious enough. There is really no alternative pathway for organized movement to emerge when the parameters of limb growth and muscle or tendon lengths are in constant flux. Even the neural control system is constantly changing, with early movement being spinally organized and later motion coming under brain control thanks to advancing myelination and general maturation of the central nervous system. The empirical evidence indicates that the notion of genetically determined motor programs is fictitious. These simply do not exist in the real world of empirical science.

We do not know of any way in which genes *could* code for behavior in this way. The problem is just too complex for the limited information storage capacity of the genotype. Even relatively simple, species-typical movements do not develop normally in the absence of relevant sensory input, as illustrated earlier by the flight responses of moose and the species-typical call of the herring gull chick.

If genetic determination is a nonstarter in respect to specific movements, what about the possibility, highlighted by many evolutionary psychologists, that what is inherited biologically is a more general psychological disposition that affects how we behave by influencing what we feel? Since gender differences in psychology were a major stimulus in the emergence of evolutionary psychology, that seems like a reasonable place to begin.

GENDER DEVELOPMENT

The notion that individual destiny is written in genes is fallible. This fact is well illustrated by development of the motor system where the type of locomotion preferred is very much a product of incoming sensory information (that is admittedly affected by anatomical development). Even skeletal development is itself caught up in a random developmental walk. Exposure to pathogens or toxins slow bone growth and increase bodily asymmetry, for instance.

In nature there is a surprising range of developmental options concerning sexual behavior and even sex organs. This fact is very much relevant to human sexual development and gender identity.

The range of natural variation in sexuality is staggering and defies the stereotype of a world neatly divided into two genders as determined by the presence or absence of a Y-chromosome. Bacteria are not gendered at all, and sexual reproduction consists of an exchange of genetic material between two individuals that may go in either direction. Fish run the gamut of sexual possibilities. Many species determine gender via dominance with the most dominant individual functioning as the male of a group.

Among mammals, the development of female offspring can be affected by male siblings with whom they share the womb. Androgens produced by the brother bleed across the placenta and produce masculinizing effects on her brain. Freemartin sheep are characterized by increased aggression that may affect their capacity to dominate other ewes if there is conflict over food or a favorable resting place. Such natural variation among females is found for other species and may have little obvious effect on fertility or health, as opposed to more extreme masculinizing phenomena where some females direct masculine sexual behavior, such as mounting, toward other females.

For some species, androgenizing effects alter external genitalia whether this is normal for that species or not. Wild bears have been seen with complete female and male genitalia. This phenomenon defies the general rule of sexual differentiation in mammals according to which the primordial sexual structures (Mullerian duct) develop either in the male or the female direction.[34] Reproductive females have been found with a baculum (or penis bone) that is characteristic of male bears. In one case, a female, whose external genitalia were replaced with a penis, gave birth to a cub, indicating that she had given birth through the ureter, hard as that is to imagine.

Androgeny is taken to a whole new level among hyenas. All of the females are equipped with a pseudopenis that is of equal size to the male penis, which is to say, unusually large, although it differs in shape at the tip being somewhat broader, permitting penetration by the male organ during copulation. Among hyenas, females are larger than males and also highly aggressive and active in dominance disputes. As in some other social canids, such as wolves, the dominant female suppresses reproduction in subordinates. The dominant female has a large litter that is raised and protected thanks to the efforts of the entire group, which includes regurgitated food carried for long distances.[35] The penis or pseudopenis is an important organ in social display, being prominent in greeting ceremonies wherein subordinates kowtow to dominants. During such social interactions, the penis or pseudopenis is erect, signaling dominance.

The peculiar reproductive system of hyenas may help explain why the pseudopenis is so consequential. Its importance for overall breeding success is underlined by the great initial reproductive cost of having this communicative organ. To be born, pups must travel through the long pseudopenis while their oxygen supply is limited or cut off. As many as 60 percent of first births are stillborn.[36]

Despite the ambiguity of gender development among some animals, including bears and hyenas, sexual development of mammals is mostly cut and dried and in accordance with chromosomal gender. Among fish, matters are very different because a given individual can have the potential to be either male or female depending upon social circumstances. For many species of wrasses and groupers, the male is the socially dominant individual. After a male dies, the socially dominant female is transformed into a phenotypic male.

Gender fluidity is taken to its logical extreme by tobacco fish, where members of a mated pair alternate between the relatively costly service of egg laying and the less costly service of fertilization. So a female who lays the eggs in the morning may fertilize the spawn in the evening.[37]

Although mammals are considerably less diverse in their gender development pathways, there can be considerable ambiguity in gender development thanks to the diverse biochemical pathways affecting somatic and behavioral gender. Androgen insensitivity is one such phenomenon, and its impact varies in degree depending upon whether it is complete or partial. Androgen insensitivity is caused by faulty androgen receptors due to genetic mutations. Where testosterone has no effect on the receptors, the condition is referred to as complete androgen insensitivity.

In complete insensitivity, genetic males (despite having XY sex chromosomes) look and behave like heterosexual women. Androgen insensitivity is often highlighted as an instance of the unreliability of genetic determination given that genetic sex determination plays second fiddle to hormonal effects.

For genetic males with androgen insensitivity, the message of increased testosterone is sent out, but the body does not respond due to nonfunctioning receptors. As genetic males, they produce the sex-determining region y protein that instructs the body to grow testes. These also produce testosterone that circulates through the bloodstream. Given that the testosterone receptor is not working, the hormone has no effect on target tissues. In the absence of instructions to the contrary, the individual develops a stereotypically feminine body shape thanks to estrogens from the adrenal gland and circulating androgens that are converted into estrogens.

So the body of a genetic male develops as though it were female—at least so far as external appearance is concerned. There is a vagina with labia and clitoris, although it is usually shallower than normal. With testicles—usually hidden deep in the body cavity—and without ovaries, the androgen insensitive male cannot have children as a woman. The testicles are often surgically removed because they pose a cancer risk.

So the androgen insensitive man looks exactly like a normal woman in terms of external appearance. What of psychology and behavior? Given that androgen receptors in the brain do not work, the brain is never masculinized. The individual thinks and feels as a woman and is accepted as such by other women.

The vast majority of androgen insensitive males are sexually attracted to men. They dress like women, have similar interests as most women, and pursue occupations that attract women while avoiding the riskier occupations and leisure activities that appeal (or used to appeal) disproportionately to men.[38]

It is remarkable that a person's body develops so differently from their genetic sex, thereby illustrating the limits of simple-minded genetic determinism. Yet the example could be turned on its head. After all, a gene affecting testosterone receptors influences not just bodily development but also complex behavioral predispositions ranging from gender identity and sexual orientation to risk taking, occupational choice, manual dexterity, sociability, and extroversion and all of the other empirically based differences between men and women.

Using complete androgen insensitivity to argue for the power of genetic determinism comes with a caveat, however, and it is a huge one. The condition is a disease that impairs normal sexual development and reproductive function. It is also selected against because those affected cannot have children.

Like most other genetic disorders, complete androgen insensitivity is extremely rare (involving about one in 130,000 individuals). Natural selection keeps the gene at very low levels. As to why the gene persists at all, this is a mystery shared by other genetic disorders. Mostly, the gene is presumed to have some advantage for carriers although there is a cost to affected individuals (who are typically double recessives).

It is relatively easy to devise a harmful mutation that messes up a complex developmental process such as sex determination. Just because sex determination is vulnerable to any number of point mutations, it is wrong to claim that any one of the affected genes accounts for a person's sex. Each of them is like a player in a very large orchestra that gets noticed only when it goes out of tune or breaks a string. Androgen insensitivity is affected by well over four hundred different mutations.[39] Identical mutations may produce a range of phenotypes, further complicating any claim of genetic determination.

Just as the biology of gender development in the human male is vulnerable to phenotypic variability, there are several pathways through which genetic females may be masculinized, either in genital or internal reproductive organs or in behavior. One of the most well-known and best understood of these is congenital adrenal hyperplasia (CAH), or adrenogenital syndrome, in which the mother's overproduction of testosterone masculinizes daughters in the womb. Although frequently detected first in the daughter, this is actually a genetic disorder of the mother where the adrenal gland produces too many cells (hyperplasia). Since the cells normally produce a small amount of testosterone, the mother's bloodstream carries atypically high levels of androgen, and this may have partial masculinizing effects on her daughters but have no detectable impact on sons, who have functional testes that produce their own testosterone from nine weeks gestational age.

Adrenogenital syndrome can masculinize the female brain. Effects are probabilistic rather than determinative but they offer intriguing insights about how the prenatal environment can shape psychological development. Admittedly, girls with adrenogenital syndrome produce more testosterone from their own adrenal glands, but the effects of CAH resemble those of a birth control pill, diethylstilbesterol, mistakenly continued during pregnancy because the mothers were unaware of their condition.[40] In each case, there is a significantly increased probability of daughters having a lesbian sexual orientation. Moreover, there is an increased interest in male-typical toys, such as cars and guns, and diminished interest in female-typical ones, such as dolls. Currently, we do not have a sufficiently well-developed grasp of brain development to understand exactly how these gender differences are produced in the young brain, except that they are mediated by the levels of circulating hormones in the fetus. Although homosexual orientation runs in families, and thus has a clear genetic influence, it is also affected by the prenatal environment. Genes predisposing people to homosexuality are presumably selected because they have multiple effects and provide advantages unrelated to sexual orientation as such. Female carriers of genes linked to male homosexuality may have higher reproductive success.[41]

In general, when we find that some trait is highly heritable based on twin or family studies, we assume that the relevant genes must provide a selective advantage. Unfortunately, genes have so many possible effects in the protein soup of brain development that the selected genes may provide advantages unrelated to the trait being studied. So the conclusion that personality traits are heritable is far less informative than it might at first seem. The problem, of course, is that a given personality trait may be affected by large numbers of genes whose contributions to the trait are extremely difficult to tease out. This point may be illustrated by the trait of extroversion.

PERSONALITY DEVELOPMENT AND EXTROVERSION

Extroversion is of great interest to gene selectionists because it ought to affect reproductive success. Extroverts are unusually active mentally and physically. They act out their impulses more and have more lifetime sex partners.[42] Extroverts are better at managing stress in modern environments although vulnerable to boredom. In the evolutionary past, they likely relieved boredom by seeking companionship and having sexual relationships. Did this improve their reproductive success? Natural selection seems strangely blind to the benefits of being an extrovert.

Some people are extreme introverts. Others are extreme extroverts. Most of us are somewhere in between. Extroversion is highly heritable, implying that it has a genetic basis. The fact that people still vary widely on the extroversion continuum implies that natural selection failed to deliver a clear verdict for extroversion over introversion.

The persistence of both introversion and extroversion in modern populations suggests that there is a balanced polymorphism. The average benefits of being an introvert would thus be roughly equivalent to the average benefits of being an extrovert so that selection could not decide between them.

The problem about this thesis is that the benefits of extroversion seem to be clearly stronger. In particular, extroverts are better at handling stress, thereby conferring a probable advantage for psychological and bodily health. Extroverts are also more sociable and outgoing so that they are in a better position to meet members of the opposite sex.[43] Moreover, they have larger social networks than introverts, who characteristically spend more time engaged in solitary pursuits. Evidence from both indigenous populations, such as the Tsimane of Bolivia, and modern societies shows that extroverts have more children than average.[44] This is a substantial advantage for men and a modest advantage for women.

If extroversion provides a clear reproductive fitness advantage, then introversion would be strongly selected against and extremely rare, yet that is not the case. Why do introverts persist in approximately equal numbers as extroverts?

The simplest explanation for the persistence of introversion is that extroversion has a downside. Extroverts may be more effective at enhancing their own social status by virtue of their extensive social networks. In the process, though, they are more likely to get into arguments and have a higher risk of violent injuries.[45] Extroverts are also at greater risk of contracting infectious diseases due to increased social contact, including sexual relationships. Their social drive might also predispose them to neglecting their own families so that their marriages crumble and their children are at greater risk of harm.

These factors must provide a real advantage for introverts, even if this is counterbalanced by the greater capacity of extroverts to handle psychological stress with associated health advantages. Extroverts have a substantial reproductive advantage, however. By that logic, genes predisposing people to extroversion should long ago have beaten out those that predispose to introversion. There should be almost no introverts today.

How can one reconcile the reproductive advantage of extroverts with this lack of evidence for natural selection reducing introversion? One possibility is that there are no "extroversion genes" on which natural selection could act.

Yes, it is true that extroversion is strongly heritable just as some other personality traits are. Oddly, this does not mean that there are genetic variants that predispose a person either to extroversion or introversion.

Thanks to the success of the Human Genome Project, it is now possible to study genotypes in detail and analyze them for correlations with personality traits using sophisticated statistical procedures. When this was done, researchers drew a complete blank for extroversion.[46] There is no evidence of particular gene variants underlying extroversion that would account for the persistence of introversion in the world despite extroverts having more children. In other words, natural selection has no relevant genes upon which to act.

So how could extroversion be so substantially heritable? Researchers suspect that extroversion is more of a side product of being strong and healthy than some trait that may be inherited like eye color.

How can you have high heritability for extroversion without genetic alleles that distinguish introverts from extroverts? One hypothesis is that social assertiveness is dependent on being in good health. It is a facultative adaptation that gets expressed only if a person is in good condition, meaning that they are strong, well-nourished, and physically attractive.

This principle likely holds for all social animals: those that are in very poor health avoid social interactions. Farmers know that a sheep who is separated from the flock is likely to be sick. By the same reasoning, individuals who are in better health are more actively engaged in social interactions and more prepared to assert their dominance over others.

Humans who are strong, healthy, and physically attractive are also more likely to enter the rough and tumble of social interactions with strangers. Extroverts are generally more physically imposing and more physically attractive than introverts both in modern societies and among indigenous residents of the Amazon region.[47] They are better equipped to assert themselves in relationships with others, even if this entails risks of conflict and aggression.

Evidently, extroversion is not associated with specific genes that are subject to natural selection. Instead, what is inherited are the thousands of genes affecting disease resistance, growth, physical attractiveness, and strength. These traits contribute to being in good physical condition. People in good physical condition behave in more extroverted ways.

If this condition-dependent hypothesis is correct, it forces us to think of personality as a lot more fluid than was true before. One implication is that a person who overcomes serious illness is liable to behave in more extroverted ways. Likewise, the strongest child in a group of relative weaklings may be more socially assertive than if they were in the company of athletes.

The complexity of behavioral development is certainly challenging, but embracing that challenge is rewarding. We grasp the real business of evolution, as it crafts phenotypes that can survive and reproduce in a competitive and sometimes hostile world. Evolutionary selection is not beholden to genes alone but uses each of the tools in its toolkit, which is to say the full range of developmental and behavioral mechanisms. If we take that perspective seriously, and correctly place humans as products of the evolutionary process, then it follows that, in their struggle to succeed, each species must be very well suited to the way it makes a living. An orangutan must be far more successful in its own habitat than a human would be if challenged to survive in the same setting. That experiment has never been tried, so far as I know, despite the close observational work of Jane Goodall types. Nevertheless, humans have challenged apes by raising them in human homes to, in effect, ask how good they are at being human. Of course, such experiments are not investigations into ecology so much as a time-honored quest to show that humans are superior to the rest of nature, and therefore not really a part of nature at all.

I do not agree with this entire line of reasoning, but it is better to give opposing viewpoints a fair hearing. If we look at the objective evidence with the impartial eye of an imaginary biologist from another galaxy, do humans really stand out as being qualitatively different from other great apes? Are we more intelligent, more removed from nature, more creative, more empathic, more skilled, more adept at communicating, or better at solving problems? The answers to all of these questions would have seemed obvious a century ago. As we have learned more about other species, the conclusions on each of these questions, and many others, have become a great deal more uncertain.

4

HOW DIFFERENT ARE HUMANS, OBJECTIVELY SPEAKING?

The outstanding achievements of human societies are mostly very recent. If we were to travel back in time a million years, we would find a species that lived self-sufficiently in nature—taking everything necessary for life out of the immediate environment and not leaving much of a mark as they moved from one part of their range to another. Indeed, this remains true of a handful of indigenous people who live mostly out of contact with modern societies, such as the Machiguenga of Peru. A million years is a short time in evolutionary terms. Yet, apart from the possible use of fire, our ancestors lived a lifestyle not materially different from that of chimpanzees who supplement their vegetable diet through occasional hunting and use simple stone tools to process their food, such as using a hammer and anvil to crack large nuts.

Charles Darwin was impressed by the close anatomical similarity between humans and chimpanzees and noted that the resemblance was particularly striking if one compared the anatomy of a young chimpanzee and a person. The close genetic relatedness of humans and chimpanzees confirms Darwin's hunch about the human evolutionary tree deriving from apes. (Of course, that impression was confirmed by modern genetic comparisons revealing very high levels of shared DNA.)

Clearly, we are all products of the same evolutionary process. That conclusion became obvious to Darwin, much as it disturbed him. Subsequent scholars struggled—and are still struggling—to come to terms with the implications. In the past, the uniqueness issue revolved around what we could do as individuals. Today, with the emergence of evolutionary approaches to social learning, the focus shifts to what we can do in groups.

Do humans, like other species, respond to environmental differences in ways that generally help us to survive and reproduce? Among birds, for example, polygyny is an adaptive response to endemic diseases and parasites. Females preferentially mate with disease-resistant males who advertise their health by means of bright plumage displays. If birds adopt polygyny as a defense against diseases and parasites, for example, can we expect human polygyny (i.e., polygamous marriage) to be more common in countries where health risks are greater?

To many readers, this question will seem preposterous. How is it possible to compare the mating of beasts in the field with the complex and refined social institution of marriage? The answer is that marriage, however dressed up in ceremony, ritual, economic contracts, laws, and conventions, is also a breeding system and therefore subject to natural selection. If human behavior is shaped by evolution (which, remember, has a lot of other tools at its disposal apart from just genes), then the form of marriage would adjust to environmental differences in much the same way that animal breeding systems do.

If we were unique, perhaps the evolutionary mechanisms shaping other species are irrelevant to us. If so, then it would really be foolish to suggest that human marriage is affected by environmental differences in much the same way as animal breeding systems are.

ARE HUMANS UNIQUE?

In school, many of us learned that humans are the crowning glory of creation, beings created in the image of God and destined to prevail over the animals and to bend nature to our will. The assumption of human uniqueness, or exceptionalism, shaped everything that we learned not just about ourselves but what we learned of animal behavior and evolution. Other species became a foil for our own gem-like brilliance. That narcissistic self-image has taken a few dents in recent decades. As animal behaviorists tested the supposed indexes of human superiority, they came away chastened. For example, our spatial memory is puny compared to that of Clark's nutcrackers and chimpanzees. [1]

The theological notion that humans occupy a special place in creation posed a problem for Charles Darwin, who recognized that humans bore many striking similarities in anatomy to chimpanzees. [2] Indeed, we are so similar to chimpanzees in anatomy and genetics, that our friendly biologist from outer space, who saw these matters dispassionately, would have no trouble placing us in the same genus. Such a close phylogenetic relationship is quite inconsistent with the notion of a special creation for humans. Darwin suspected that chimpanzees and humans are closely related species that might both be derived from a common ape ancestor.

In addition to his interest in anatomy, Darwin was a keen observer of animal behavior. He was impressed that chimpanzees would console another who was upset by wrapping an arm around them and saw in this reaction a suspiciously human-like capacity for empathy. Empathy is not all they are capable of, of course, and zookeepers know that chimps are strong, dangerous, and unpredictable creatures. Contemporary scientists investigated a number of their less cuddly activities. Chimpanzees in the wild sometimes hunt monkeys and other animals for food. Male chimpanzees practice infanticide, presumably as an unconscious way of ensuring that females invest more in their own offspring. Rival bands of males take each other on in territorial warfare where a band of males gangs up on a vulnerable member of another group and kills him, presaging human political alliances and warfare.[3] Nasty as these practices are, they are reported by anthropologists studying some preindustrial human societies. Warfare continues in modern societies, of course, although infanticide (if not medical abortion) is criminalized.

The terms of the human uniqueness debate in animal behavior echo the old Darwinian controversy about human special creation versus descent from the apes. Strangely, the rhetorical structure of this debate persisted long after scientists accepted Darwin's conclusion that we are descended from apes. Primate researchers still look for qualitative differences between humans and apes. Are humans capable of some feat of intelligence or communication that is completely beyond the capability of chimpanzees or other animals?

This investigation may seem scientifically pointless if one accepts that chimpanzees and humans are in the same lineage. A close common ancestry automatically implies that we have a great deal in common, behaviorally and cognitively, as well as anatomically, although it does not close the door on conspicuous psychological differences either.

THE AHA MOMENT

Austrian behavioral scientist Wolfgang Kohler waited out World War I on the island of Tenerife in the Canary Islands, where he amused himself by conducting experiments on chimpanzees. Kohler wanted to know whether chimpanzees could solve abstract problems as humans do.[4] His most famous problem was the banana-and-box experiment. A banana would be suspended well out of the chimpanzee's reach with a large packing crate positioned at a good distance from the banana. A correct solution required the animal to drag the crate until it was positioned below the banana and then climb on top of it so as to reach the prize.

When first presented with this problem, the chimpanzee tried leaping at the banana, but it remained out of reach. At this point, the animal might give vent to a fit of irritation and redouble its efforts, all in vain. Then it would wander away and become quiet, as though deep in thought. Suddenly, the chimp would get up, grab the box, and drag it beneath the suspended banana. By climbing on the box, it succeeded in grasping the banana. When presented with the same problem on subsequent occasions, the chimpanzee solved the problem without hesitation.

Kohler referred to problem solving of this sort as insight. Like a human faced with a difficult puzzle, the animal focuses its mental efforts on finding a solution. Eventually, the correct solution comes to mind. This is the "aha moment" when the problem is solved mentally. All that remains is to put that insightful solution into practice. For Kohler, a key consideration in insight is that once a problem is solved, it stays solved: there is no need to think about it again.

Subsequent researchers questioned Kohler's inferences about what may or may not be going on inside the chimpanzee's mind, yet there is no doubt that chimpanzees can solve abstract problems and that they learn to solve human problems from associating with people. They know that you won't get water out of a hose unless you turn on the tap and that a fan won't work if it is not plugged in. If chimps can solve such problems almost as well as humans, perhaps they solve them in much the same way as we do.

That cognitive interpretation did not sit well with behavioral psychologists in the tradition of B. F. Skinner, who argued that if you cannot observe what is going on in the animal mind, you are better off not discussing it at all. He preferred to demonstrate that animal behavior could be molded by events external to the creature. Pushing the box toward the banana was reinforced by getting the fruit and eating it. Skinner's academic descendants, including Robert Epstein, worked to show that what looks like intelligent behavior in chimpanzees, or other animals, might be explained instead as behavioral reinforcement. [5]

Epstein trained pigeons to push a light box underneath a plastic banana that they pecked to receive a food reward. Superficially, this looked like what Kohler's chimpanzees had done in the banana-and-box test. Yet it was the opposite of intelligence because pigeons were reinforced separately for pecking the plastic banana and for pushing the box toward it. This technique is known as "shaping," and it is used by animal trainers to create circus tricks. Instead of convincing anyone that pigeons were insightful problem solvers, Epstein had merely demonstrated how skilled he was as an animal trainer. What looked like insight could be produced by shaping up each of its components separately and giving the pigeons an opportunity to combine them, which they might well do if they became confused about which task they were getting "paid" for.

Much as the behaviorists objected to considering animals as humans (or anthropomorphism), Kohler's perspective won out. Animal behaviorists now accept that chimpanzees solve abstract problems in their heads much as we do, implying an inner mental life. Humans may be more intelligent than other animals, but it is only a matter of degree, and some animals exceed our abilities in specific cognitive skills, such as spatial memory. We are not unique in our capacity to reason as scholars had assumed before Kohler.

The role of intelligence in solving practical problems is easy to overestimate. Even very unintelligent animals, such as honeybees, control their environments in remarkable ways. These animals protect themselves from temperature extremes by constructing a hive to live in. The temperature inside the hive is regulated to within one degree centigrade, which is better than most human HVAC systems can do.

Whatever the marvels of social organization among honeybees and other social animals, are they very limited in what they can communicate with each other compared to the richness and complexity of human language?

LANGUAGE IN OTHER SPECIES?

Are humans unique in their capacity to learn a language? Given that chimpanzees are so intelligent and use hand gestures naturally (such as a cupped hand used to beg for food), might they be trained to use American sign language? If so, then we might open up a frontier of cross-species communication across which we might come to know animal minds.

This lofty goal began with more mundane objectives. Chimpanzees were trained to make simple requests, whether for a food item or to play with the trainer.[6] This research seemed to go well in different labs, and star pupils evidently mastered more than a hundred distinct signs. Yet there were problems. Chimpanzees had appalling grammar and seemed insensitive to basic semantic rules, such as the correct ordering of words. They used the signs to construct strangely inelegant sentences such as "Trainer give me banana banana banana me give."

The chimpanzee language experiments were bedeviled by a lack of objectivity. Trainers were biased in wanting to believe that their subjects were making progress and using symbols correctly. When the chimpanzees were tested by strangers, using rigorous scientific procedures, their supposed language ability crumbled.[7] They were doing little better than chance. In other words, they were incapable of communicating meaningfully using sign language.

Evidently, trainers had unconsciously given cues to the chimpanzees. In many instances, the subjects merely repeated signs that they had just witnessed. This was not the end of the ape language story. The sobering findings convinced scientists that they needed to use better protocols that pushed the trainer into the background. Ultimately, chimpanzees learned to push buttons on a computer keyboard indicating that they could connect a symbol with its meaning. One of the star pupils was Kanzi, a bonobo (or "pygmy" chimpanzee) who learned many signs simply by accompanying his foster mother to training sessions. In addition to mastering more than four hundred words, Kanzi had good comprehension of complex sentences formed by others, though his own utterances were brief, mainly consisting of one or two symbols that communicated requests for food, toys, or activities.[8] Bonobos may have made good subjects for language training because they are hypersocial under natural conditions whereas common chimpanzees seem more independent and express their sociability in more Machiavellian ways.

Common chimpanzees made poor linguists, but they are extremely intelligent, nonetheless. The first convincing evidence of language capability emerged from far less intelligent animals, namely honeybees.

LANGUAGE BEYOND THE APES

Honeybee scouts leave the hive each day in search of good patches of blooming plants from which nectar is harvested. Once they find something promising, scouts return to the hive and inform the other bees about its direction and distance.[9] This information is communicated in a dance performed inside the hive known as the waggle dance. Remarkably, the dancing scout bee informs other workers about the direction of the find by transposing the angle made by the food patch and the sun in relation to the hive to an angle from the vertical that becomes the axis for the waggle dance. The amount of food available is communicated by the speed of the dance. The scout may tell other workers about the direction and distance of the foraging site, but how is this language? It is language because it uses arbitrary symbols to depict objects that are not present.

Most linguists would accept the waggle dance as a rudimentary language communication system. Yet there is no cottage industry of young graduate students lining up to teach bees sign language. The bee language has only about three words, and they always refer to a food patch, denoting its direction, its distance (given by the duration of the dance), and its amount. That is all that a bee can communicate in its "language."

Bees and other social insects have many remarkable qualities, but their language tells us next to nothing about the evolution of human language. We would be well advised to stick to more closely related animals that at least have a backbone. Birds, in particular, have an elaborate communication system that may not be a true language but nevertheless offers many remarkable analogies with human speech.

Male birds sing during the breeding season when their testosterone level rises and primes both the syrinx (or voice box) and the brain to promote singing. Females that are treated with testosterone in an experiment also develop the capacity to sing. Birdsong has two primary functions. The first is territorial: it informs rivals that the territory is taken and that any intruder can expect a beating. This is not an exaggeration because a territorial male is highly confident. He has the home turf advantage and is most likely to win an encounter with a rival the closer he is to the center of the territory. The second major function of birdsong is sexual advertisement. A healthy bird sings with a stronger voice and can sing more loudly and for longer. One that is ill sounds weaker and tires more quickly so that it sings less loudly and spends less time in song. Females detect these cues and avoid mates who sound less healthy because a sickly mate transmits its vulnerability to disease and parasites to the offspring. So a male with a vigorous song is more sexually attractive because it can transmit disease resistance to offspring.

Important as a male's song may be in practical terms, the territorial song directed to other males transmits limited information. To rival males, it says: "This territory is taken and will be defended." The territorial song is generally simple and short, but it gets repeated ad infinitum. Some species sing courtship songs that are different from the territorial song and generally more elaborate and varied. To an interested female, the courtship song essentially says, "I am healthy and virile." Despite the simplicity of what is communicated, the more information researchers collect about birdsong, the more interesting it becomes for two good reasons, First, birds communicate a lot more information in their song than a casual observer could detect. Secondly, birdsong has many analogies to human speech both in terms of how it develops in the lifetime of the individual and in how it varies across populations.

Some birds, including mockingbirds, sing the songs of other species and can even imitate mechanical sounds such as the noise made by a teakettle or the ringing of a telephone.[10] Why they do this is not entirely clear, but there are several reasonable explanations. One is that females are attracted to males by their singing virtuosity that signifies a superior nervous system. A bird that masters many different songs may be good at various survival skills, such as finding food or evading predators. So it is rather like women being attracted to signs of intelligence in men that predict occupational success.

An entirely different line of reasoning is known as the Beau Geste hypothesis after the story of a soldier who successfully defended a fort by propping up his dead companions around the parapets and running around to fire from their positions to convey the impression that the place was well defended thereby inhibiting the attackers. According to the Beau Geste hypothesis, a bird sings many different songs to convince new arrivals that the territory contains a lot of birds already and is thus well defended, encouraging the rival male to try his luck elsewhere. The singing of songs from different species may reduce competition with other species in addition to the singer's own.

So birdsong may convey a great deal more information than is contained in the trills and runs of the physical song. It can tell a female about the health and intelligence of the singer. It may also repel rival males whether by communicating strength and confidence or by sending deceptive communications about the number of birds in the vicinity. Yet that only scratches the surface of what can be conveyed in a bird's song.

One of the more intriguing aspects of the analogy between birdsong and human language concerns local dialects. In indigenous societies, people growing up in one local area may speak very differently from their neighbors ten miles over the mountain. Even if they speak the same language, they use recognizably different pronunciation and phrasing. Such differences are most elegantly explained as copying errors. When each generation learns to speak, they copy the speech patterns of their parents and other adults around. The learning process generally produces something very close to what was heard, but it is not perfect. Such random change from generation to generation gets magnified over time, generating quite noticeable dialect shifts between one village and its neighbors just ten or twenty miles away. Because they communicate most with people who live nearby, a certain uniformity is maintained within that language community. If it were not, speakers could not understand each other.

No such corrective process applies between villages twenty miles apart whose residents rarely communicate with each other. So accumulated errors get carried on, and potentially magnified, from generation to generation. These local dialects were based on linguistic isolation in a world where many people never traveled more than ten miles from their native village. Different dialects shaded away into different languages that were mutually unintelligible. So an undeveloped island like New Guinea, which had difficult terrain and no paved roads, maintained linguistic diversity with perhaps more than seven hundred distinct languages.[11]

In New Guinea, linguists found marked shifts in dialect every twenty miles or so, and these shifts are attributable to linguistic isolation of local groups, an isolation that is reinforced both by geographic barriers, such as rivers and mountains covered in tropical vegetation, and also by tribal affiliations with each tribe jealously guarding its home turf.[12] New Guinea's songbirds were also found to have dialect shifts, and the same is true of birds elsewhere. This includes American birds such as the white-crowned sparrow whose song is widely studied by researchers. Dialect shifts are sometimes observed over distances as small as two miles.

It might seem odd that birdsong dialects would be contained in such a small area given that birds can fly ten times that far in less than an hour. The existence of dialects on this geographic scale indicates that some songbirds have strong fidelity to their place of birth. If they moved around more, then local dialects would disappear. This means that when young birds separate from their parents, males establish territories close to that of their father.

Females of most species do not sing, yet they are affected by song dialects. For example, females of some species may prefer the song of males who sound like their fathers. In other words, they prefer males who sing in the local dialect. This could be an advantage if they breed in a place that is already somewhat familiar where they can more easily find food or get shelter from a predator. It also means that the song contains reliable information about where a male is from that is intelligible to potential mates. Females of some species *avoid* males that sound like their fathers, however, possibly to prevent inbreeding.

Birds learn to sing in ways that have both striking similarities and striking differences from the way that children acquire language. When children learn to speak, they begin by babbling: basic speech sounds are uttered repetitively: pa-pa-pa, da-da-da, and so forth. This phase begins at around five to six months, and by the time an infant reaches eleven to twelve months, it is uttering its first words. If an infant does not hear spoken language during this period, it has trouble learning to speak normally, as illustrated by isolates such as Kaspar Hauser, who spent his childhood locked up and socially isolated. So there is a sensitive period for language development during which the brain is especially receptive to speech sounds. Over the next several years, a child masters the complex rules of grammar and syntax. By the age of five years, most children have mastered complex sentences and can decode most of what adults are saying around them.

Children learn to speak using the exact pronunciation and intonation of those around them. A careful listener can always hear the difference between a native speaker of the language and one who acquired it later in life. Immigrants often persist in speaking with a heavy foreign accent even after they master all other aspects of the new tongue. It is as though their accent were fixed early in life by their first language.

Some of these peculiar features of language learning are mirrored in the way that birds learn to sing. Before singing, young birds go through a repertoire of song syllables that are uttered in a disjointed way and have no communicative value. This sub-song resembles babbling in infants. Birds also have a sensitive period as nestlings when they must hear the song of a male, typically the father, if they are to master the adult song. It is as though a window opens early in life when the brain records the sound of its father's song. Much later when a bird begins to sing, it matches its own utterances with the recorded song, or template. In this way, an individual produces a song that matches that of its father. This astonishing sequence was pieced together through careful experiments by Peter Marler and others, and it helps explain how a bird may learn to sing at a time of the year when other adults are not singing around it (because the breeding season is over).[13]

We are accustomed to thinking of birdsong as a repetitive short phrase, but this may be selling birds short. We begin to appreciate what birds are capable of when animals like parrots give our own sentences back to us. Most observers write this off as a sort of mindless repetition, but Irene Pepperberg has shown that her parrot, Alex, can learn to respond in meaningful ways to questions she poses to him.[14]

Of course, such experiments are vulnerable to the problems of interpretation raised in the ape language controversy. One person's meaningful answer is another person's conditioned response. For example, a parrot can be reinforced to say "green" in the presence of a green object without having any understanding of the semantic relationship between the sound and the color. Parrots are good at repeating sounds, and they are relatively easy to train because they are so highly social as reflected in the deep attachment between an individual and its mate that often gets transferred to an owner.

Given that birds are primarily songsters, it is interesting that some species can mimic all sorts of sounds, from a telephone ringing (in the case of American mockingbirds) to a shepherd whistling to his sheepdog (European blackbirds) or chainsaws and camera shutters (Australian mynah birds). Peter Marler and Paul Mundinger describe the work of ecologist Erwin Tretzel, who went beyond mere collection of anecdotes to produce high-quality recordings that could be scientifically analyzed. His results were published in German scientific journals:

Although there are many anecdotes about interspecific mimicry in nature, few examples have been submitted to acoustical analysis. Tretzel has made this subject his special interest.

A crested lark (*Galerida cristata*) in Bavaria has been found imitating from different whistle commands given by a shepherd to his dog. . . . The dog responded correctly to tape recordings of the crested lark imitations. Similar, though less perfect imitations were detected in the songs of another lark 3 km away, presumed to have learned its version from the first bird.

In another part of Bavaria, Tretzel discovered a group of European blackbirds (*Turdus merula*) singing what appeared to him to be an imitation of human whistling. Some of the imitations were more precise than others. By tracking them down, he was led to the house of a man who had been in the habit of whistling this tune to his 8-year-old cat since it was a kitten. Analysis of recordings of his whistling revealed he was no musician and the pitch and rhythm varied considerably. The blackbirds transposed the entire motif to a higher pitch, with some additional notes, and maintained it in a more regular fashion than the man's original versions [citations omitted].[15]

So far, birds' natural songs were discussed as simplistic refrains that get regurgitated with some error from generation to generation. In some cases, they are far more. In large populations of birds, the song can grow quite complex and lengthy. It is rather like a human musician who changes her performance in response to what other artists in the community are doing. This phenomenon is illustrated by what happens with European meadow birds whose habitat is disrupted by plowing. Sometimes what had been a large community of Dupont's larks gets diced up into isolated "islands" of rough land where the tractors cannot go.[16] Researchers found that among the isolated communities, their song declines in complexity as resident males get separated from virtuoso singers and are reduced to copying the humdrum efforts of the average Joe instead.

This scenario has all kinds of intriguing implications. To begin with, song for birds may be a competitive endeavor, much as it can be for humans: performers try to outshine each other. This implies that birds are striving to impress an audience. The audience could be rival males, but there is no particular reason why rivals would be impressed by the fine turning of a musical phrase. After all, rivals would be primarily interested in the strength and stamina of the singer as an index of fighting ability, and such traits are better represented in song loudness and duration. So it seems more likely that the song complexity is an effort to charm the females, as Darwin expressed it.

Just as many women are attracted by men who are good conversationalists, it appears that female birds are attracted to males whose song complexity stands out. Why would they care? One possibility is that complex songs are a test of brain function: males who succeed in this musical test are high functioning and healthy and thus better equipped to find food or elude predators. Females drawn to such mates would derive superior genes for their offspring. What is more, their sons would be better at attracting mates after they had matured, an idea known as the "sexy son hypothesis."

Whatever the reasons that females might be attracted to the singer of a more complex song, song traditions in some bird communities bear striking similarities to human musical traditions that flourish in settings where several talented musicians can come together and challenge each other. This phenomenon reaches far beyond entertainment: it applies to human technology as much as to art and entertainment. In all cases, larger communities benefit from having more talent. This is one possible reason that immigrants are drawn to large bustling cities rather than rural backwaters.

The general principle is that smaller communities are simpler whereas larger communities foster more complex traditions whether it is birdsong, human musical entertainment, or advanced science and technologies. The importance of community size as a factor in behavioral complexity is highlighted by a strange loss of boat-building abilities in a subsistence society that draws much of its sustenance from sea food. Archaeologists discovered that Tasmanian islanders used to build sophisticated seagoing canoes but mysteriously lost that technology over time. How or why this might happen puzzled anthropologists. Many pointed to the Tasmanians as a "sick" society that somehow lost its way and failed to produce the basic tools of survival for a maritime people.[17]

In recent decades, a more satisfactory explanation has surfaced that is essentially the same as the loss of song complexity of birds in European meadow islands.[18] The key similarity is a reduction in the size of the community. The bird communities were fragmented by plowing up of habitat, effectively reducing the size of the local community.

Something quite similar occurred for the Tasmanians. They originally lived on an isthmus. Then the narrow neck of the isthmus got washed away, turning them into an island community. When they were connected with the mainland, they interacted with a much larger community. The geographic isolation of an island greatly reduced the number of people they could interact with on a regular basis. Loss of contact with mainland residents isolated the community from expert boat builders.

Such expertise likely passed from father to son. Boat-building tradition on the island hung by a thread of the few experts living there. Eventually, all either failed to reproduce or were lost at sea, so that their expert craftsmanship died out with them.

The more that researchers learn about birdsong, the more they realize that there is a lot left to know. No one really knows the extent of information communicated by birds in either their courtship songs or their territorial songs (that may or may not be different depending on the species). We do know that it has an endless capacity to surprise us. Birdsong is probably not a language comparable to human language with its thousands of meaningful words and its elaborate grammatical structure. Even so, there are so many fascinating parallels between birdsong and human language that our confidence in claiming human language makes us unique among the evolved species on this planet gets shaken. After all, if a pet parrot says, "Someone's here" when it hears a sound at the front door, who are we to say it doesn't know what it's talking about?!

Vocal communication in dolphins and whales (or cetaceans) is probably a lot more complex than bird songs, if only because these animals have extremely large brains and appear to be highly intelligent. Cetaceans' communication is poorly understood because they live in water and their vocalizations are inaudible to humans unless we go diving or place expensive sonar recording equipment in the water. Much of our current interest in whale vocalizations stems from songs recorded by the U.S. Navy who used underwater recording equipment to spy on foreign navies.

COMMUNICATION IN WHALES AND DOLPHINS

Whales and dolphins are highly social, and many have a complex repertoire of acoustic signals, the meaning and significance of which are largely unknown. Bottle-nosed dolphins are a popular animal at zoos and aquariums. They are highly trainable and put on an enthusiastic acrobatic display that pleases crowds. Dolphins, and related species, emit a series of loud clicks that are used in echolocation, permitting them to sense objects in muddy water, whether obstacles such as rocks or prey animals. These clicks are not used in social communication, which employs high-pitched whistles instead. Dolphins vocalize to each other a lot, but the meaning of these "conversations" remains to be decoded.

Humpbacked whales are another species whose vocal communication received a great deal of attention from students of animal behavior, linguists, conservationists—even musicians. Humpbacks travel together in pods that stay synchronized by means of communicative whistles. Each individual emits a distinctive sequence of sounds that serves as a vocal signature: no two are exactly alike.

Male humpback songs can last for fifteen minutes or more. These likely play a role in courtship analogous to the elaborate songs of some birds. The song is composed of two different passages. The early sequences are repeated from one session to the next, but the later sequences are more creative. It is as though a musician first plays each of the tunes he knows well and then improvises. This phenomenon suggests language because human languages have "generativity," or originality. Every time I open my mouth, I am capable of saying not just something that I have never said before but something that no other person has ever said before either.

What does humpback improvisation mean? Is it genuinely expressive? Or is it just chatter that is used to fill a gap, rather like the verbiage that students manufacture to honor a length requirement of their term paper? We simply do not know, and that is a weak position from which to dismiss the language capacity of other species.

Human language communication seems better developed than that of other species. Yet we should be wary about concluding that this is a qualitative difference rather than a quantitative one. Birds and cetaceans communicate a great deal of useful information in their vocalizations. Much of this information likely involves mate selection: the song of the males is a sexual advertisement lauding their own excellent biological and cognitive condition.

Animal communications sometimes use a particular symbol to denote a specific object, such as the signature whistle of the humpbacked whale specifying that individual animal, or monkeys emitting a different warning for predators in the air compared to threats from the ground. This passes the test for semanticity or meaning—one of the defining characteristics of language.

"Man bites dog" has a very different meaning compared to "dog bites man." We do not know whether any animal communication system naturally possesses that type of grammatical structure, but claims have been made that trained animals are sensitive to syntax, specifically some language-trained chimpanzees and Alex the parrot.

Are animal communications entirely functional? Or do they have the expressive function of human language where people converse with each other to know the other person better? Some birds, such as starlings, are highly vocal and seem to use their everyday interactions with the environment as an opportunity for vocalizing with each other. Mozart kept a pet starling to which he was very devoted and immortalized the bird by including some of its vocalizations in a composition ("A Musical Joke"). The starling chirped away while Mozart worked on his compositions, and he enjoyed its musical companionship, which is presumably why he held a funeral for it complete with mourning clothes.[19]

While Mozart may have felt that the pet starling was communicating with him, we have no idea what, if anything, the starling was thinking about. This raises a key question about the interior lives of other species. Are some self-aware as humans are? Or are they basically hot-blooded machines designed by evolution to solve problems of survival and reproduction?

SELF-AWARENESS

How would we know if another species is self-aware? One cannot ask human-like leading questions such as: "What are you thinking about right now?" For such questions, answers like, "I am worried about the split ends in my hair" could be used as evidence that the subject was self-regarding, or self-aware, because they were concerned about how others saw them.

Researchers must devise ingenious experiments to demonstrate animal thought processes, such as Kohler's insight experiments with chimpanzees. They noticed that captive chimpanzees enjoyed clowning around in front of mirrors. The chimps would pull funny faces, inspect their teeth, or place a cabbage leaf on top of their heads, like a hat. Such antics suggest that the chimpanzees recognize the image in the mirror as themselves. If so, they must have a sense of self. How can one prove this objectively?

Gordon Gallup devised the first credible test for self-awareness.[20] He exposed chimps to a large mirror so that they could get familiar with their own image. Then he changed their appearance without their knowledge. He did this by the drastic measure of injecting them with an anesthetic. When they were out cold, he placed a dye mark on their foreheads. After they had recovered, the animals were put pack in the mirrored cage to monitor their reaction. Gallup found that chimpanzees behaved very much as humans might under similar circumstances. They used the mirror to inspect the mark, touched it with a finger, and attempted to remove it.

This basic test is still widely used except that the anesthetization was dropped as too risky. Instead, subjects are sham marked or get the sensation of being marked without the dye in a control condition. Not many animals pass the test for self-awareness. They include chimpanzees, bonobos (pygmy chimpanzees), orangutans, a couple of elephants, dolphins, humpbacked whales, and magpies.[21] Apart from the magpies, all of these are large-brained mammals. They are all highly social as well, with the exception of orangutans, which are mostly solitary as adults.

Magpies are a big surprise on this list given that they have bird brains, but they, and their relatives (the corvids, or crows), are surprisingly intelligent and pass problem-solving tests that only great apes can master. Results are mixed for gorillas and capuchin monkeys with some studies claiming that they pass the mirror test but others reporting that they fail.

Surprisingly, dogs do not pass the self-awareness test and perceive the reflection as another dog at whom they may bark. Dogs are highly intelligent and extremely social and fit right in with human households, even to the extent of voluntarily learning to recognize the meaning of human words. Anyone who saw the *60 Minutes* segment on border collies knows that these clever dogs are extremely attentive to the needs of their master. One collie had a large collection of about a thousand stuffed toys that he could retrieve on demand. "Fetch Kermit" always yielded the talkative frog character from *Sesame Street*, for instance, and whichever toy was requested, the dog retrieved it from an adjoining room.

Dog lovers complain that the mirror test favors visual animals like primates but makes it more difficult for dogs that are more focused on auditory and olfactory cues. Animal cognition researcher Mark Bekoff found his dog Jethro (a neutered Rottweiler mix) could recognize his own scent marks from urine in snow but that is not exactly self-awareness.[22] It could be that most scent-marking mammals avoid overmarking their own scent in a reflexive fashion.

The bottom line is that humans are not unique in being intelligent or self-aware. Self-awareness can be considered an important cognitive Rubicon. Once crossed, all sorts of mental activity become theoretically possible. These include deliberation and planning, impulse control, deliberate deception, and Machiavellian manipulation of others. Such executive functions seem to be particularly well-developed in humans and may be facilitated by our greatly enlarged frontal cortexes relative to other large-brained species.

DELIBERATE THOUGHT AND PLANNING

Even if some species are self-aware, there is a common perception that most nonhuman species remain little more than warm-blooded machines, incapable of detailed planning, at the mercy of their impulses, unable to think deeply about their past experiences, and devoid of insight about what they know (i.e., metacognition). So planning, memory, delayed gratification, and metacognition used to be considered the exclusive features of human cognition (along with self-awareness).

Each of these prejudices has yielded in some degree to careful experiments. Anyone who works with great apes knows that these are capable of cool-headed deliberation that is sometimes described as Machiavellian, although they participate in social activities, such as giving comfort to the upset or resolving disputes. Chimpanzee males form alliances that sometimes permit a pair of allies to knock the alpha male off his top-ranking spot even though they are smaller and less physically aggressive.[23]

Chimpanzees are skilled at strategic thinking and deceptive behavior. For example, a low-ranking chimp who is sexually aroused by a receptive female will conceal his erection from the alpha male but show it to the female as a sexual invitation.[24] The male then departs into the jungle using a circuitous route. If the invitation is accepted, the female leaves in a completely different direction and circles around so as to meet up with the male. In this way, their tryst is completed without the risk of being attacked by the dominant male. Such anecdotes illustrate a clear capacity to delay gratification.

Experiments have shown that primates have good episodic memory and that they can recover buried food even if lengthy delays intervene before gaining access to the site where they had witnessed the reward being hidden. Low-ranking individuals avoid finding the food when other individuals are present: evidently, they realize that this is a sure way to lose it. So they know what they know and comprehend that others know less; in other words, they have metacognition.

The same may be true of elephants in Indonesia who are fitted with bells that help their mahouts keep track of them. They may silence the bells by stuffing them with grass so as to move around undetected. While it is possible that they merely silence the bells because they find them irritating, it is possible that they understand that humans may use the bells to monitor and control their movements.

From problem-solving corvids to political chimpanzees, virtually everywhere researchers look, they produce evidence that many species develop cognitive models of the world they inhabit. In amusing experiments with monkeys, subjects learn to work for a delicious treat, such as a grape. When they have mastered the task, the delicious grapes are suddenly replaced with boring standard monkey pellets.[25] When the expected reward fails to appear, they become very upset and refuse to eat the substitute even if they are hungry, contrary to reinforcement theory. Likewise, they are annoyed if they receive a lesser reward than other monkeys get for the same task. Clearly, they are affected by expectations about how the world is supposed to work and may even have a sense of what is fair.

An impartial reading of the evidence thus suggests that human cognition is continuous with that of other species rather than being qualitatively different. Other species store episodic memories that influence their subsequent behavior. They can plan ahead, and they show evidence of knowing what they know, as well as knowing what others know. Human uniqueness enthusiasts may still comfort themselves with the notion that humans are exceptional in their capacity to spin original products out of their own minds.

What about the human potential for creativity and artistic expression? Are we alone, or do animal Rembrandts lurk in the jungle?

CREATIVITY IN ANIMALS

The search for creativity in animals is complicated by lack of pertinent information. Highly communicative species produce complex and varied sounds that have the potential for originality. Darwin wondered whether songbirds may sing outside the breeding season purely for their own amusement, and we are still wondering about many such questions today. If a humpback whale varies its songs, is it trying to be clever to impress females in the vicinity? Does that bear any relationship to a lover serenading his lady according to Mediterranean custom? Is this genuinely creative? Or could it be that the whale varies its song to avoid injuring its vocal cords? We may never know because we have not decoded the meaning of the song. Indeed, we do not know whether it has any meaning whatever. Perhaps all that matters is keeping a female's attention for long enough for her to get in the mood for mating.

If a bird sings one hundred different songs, is it more creative than the bird who sings only ten? Such questions are impossible to answer: scientists have a limited understanding of what is being communicated and of how new songs are generated. Trying to understand what the birdsong communicates is a bit like archaeologists trying to read Maya inscriptions before they knew what any of the hieroglyphs meant. We may be unable to assess the creativity of other species in their terms, but we can certainly ask whether they can be creative in human pursuits.

Can animals produce creative products similar to those of human artists? Primate researchers provided chimpanzees and other primates with paper, crayons, pens, and other art materials to see what they might generate and how it compared to human artistic products. Early results are presented in zoologist Desmond Morris's fascinating book *The Biology of Art*.[26] The majority of primates generated little more than a jumble of random scribbles. That is hardly surprising. After all, most *people* are not very accomplished artists either. We must therefore evaluate their *best* products rather than the average output. Some chimpanzees are good painters, possibly because of their natural penchant for making and using tools that require fine manipulation of objects. Orangutans and gorillas also produced interesting paintings. A few monkeys made passable works with highly intelligent cebus monkeys doing the best.

Chimpanzee artists embraced painting with surprising passion. They would stay on task until a particular sheet of paper was more or less filled with color. After they had finished, they would refuse to do any more work, suggesting they had a sense of the painting being completed. That, of course, is a defining characteristic of most professional painters. Joseph Turner, a distinguished English painter, shocked the art world by placing a crude daub of paint across a finished work of his hanging in a gallery. Evidently, chimpanzees evince the same fear of damaging a finished work by continuing to paint over it.

Most chimpanzee pictures were not too attractive. They deployed a basic fan pattern extending from the bottom center of the page rather like a clump of trees.[27] Some worked in monochrome. Some added one or two additional colors. A chimpanzee named Congo turned out to have a lot more artistic ability than the others. He was the Rembrandt of the chimpanzee world. Indeed, if there was no Congo, chimp art would hardly have received much attention.

Desmond Morris describes an interesting test of nonhuman creativity.[28] The idea was to hang the primate paintings in an exhibition without revealing the identity of the artists. Then recruit some distinguished art critics to see what they would make of it. None of the critics suspected the work was by a nonhuman primate, or even by a child. Indeed, some believed that the anonymous paintings must have been the creation of a leading abstract expressionist like Russian artist Wassily Kandinsky. Anyone who looks at Congo's paintings can understand why they were duped. Superficially, these are blobs of color generally made in the basic fan pattern. Yet they have their own distinct aesthetic with a sensitive choice of colors and an unmistakable feeling of balance, control, and restraint. They are hauntingly beautiful.

In recent years, artistic primates were pushed off center stage by elephants. Unlike chimpanzees, the elephants made representational paintings that are comparable to the work of a talented eight-year-old child. Tourists can watch elephants make pictures in art schools dedicated to them. This provides a welcome source of income for protecting this endangered species that is no longer widely used in logging and other industries where great strength used to be valuable.

It is remarkable that elephants would not only learn how to wield a paintbrush but that they could produce representational paintings similar to what an eight-year-old child might do. Their choice of subject matter was suspect, however, implicating a human sensibility. After all, why might elephants want to generate an image of a vase of flowers? Even a portrait of an elephant seems a stretch. Do wild elephants have walls upon which to hang portraits of their friends and mentors?

Desmond Morris, who had studied the artistic efforts of primates, grew suspicious.[29] He visited an elephant painting establishment to observe the painters at work in their outdoor studio. Before long, Morris noticed a strange correlation between the movements of the elephant trunk and its paintbrush with tugging by the elephant driver (or mahout) on its ear. If the driver pulled the ear vertically, the elephant made a vertical line. When the mahout pulled its ear sideways, this was followed by a horizontal brushstroke. Moreover, a given elephant always painted exactly the same picture. Morris concluded that the paintings are really being produced by the driver using the elephant's ear as a control system. Each driver trains his beast to produce a single painting and repeats the same performance ad infinitum. Tourists who witness the elephant painting its picture are delighted to buy it for a modest sum. After all, they know what they have just seen!

Remarkable as the elephant painting phenomenon is, it is definitely not creative expression. Morris concluded that it is a mere circus trick. This is disappointing to enthusiasts for animal creativity and certainly raises the bar for anyone who now wants to claim that elephants have similar creative capacity to chimpanzees. Perhaps an elephant counterpart of Congo—the chimpanzee Rembrandt—can still emerge from the jungle. As of now, this has not happened and the fact that only one elephant passed the mirror test for self-awareness suggests it is unlikely.

Humans are not the same as any other species, but every species is unique or it would not be a species. Whatever about creativity or self-awareness, the view that humans are intellectually superior to other creatures was accepted as a truism at least since the days of Plato and Aristotle, who celebrated the potential of the human mind. Is the human brain qualitatively different from other species in ways that allow us to achieve higher-order intellectual feats? Do we fit in with other species, or stand out as unique?

WHERE HUMANS FIT IN NATURE

Humans have larger brains than other animals relative to the size of their bodies (although large cetaceans have bigger brains in absolute terms). Our brains also burn a great deal of energy even compared to our ancestral primates. Such adaptations of the brain could facilitate a wide range of intellectual accomplishments. Yet our aptitudes may not be qualitatively different from what other species can do, at least if the comparison is made in an unbiased way. Unbiased means comparing subsistence humans with other species before the advent of modern technologies. Such comparisons are impossible as a practical matter, but they can be addressed in a thought experiment. Would a human from two million years ago (a short period in evolutionary time) have seemed very different from other primates if examined by a hypothetical anthropologist from another galaxy? There are no clear lines of demarcation that would make early humans clearly different, or obviously superior, to other intelligent animals on the planet. At least, that is what contemporary research on the cognitive abilities of animals indicates. Indeed, there are some cognitive abilities where humans are put in the shade by other species.

A young male chimpanzee, Ayumu, was trained on a touchscreen to memorize a series of numbers from 1 through 9 that were displayed at random locations on the screen and were replaced by white squares as soon as he began tapping.[30] The numbers had to be tapped in the correct order. His competence at doing this was remarkable enough. What really floored researchers was that, even with extensive practice, humans could not do nearly so well on this spatial memory task if it was presented very quickly.

One does not need a large great-ape brain to be good at spatial memory either. Clark's nutcrackers hide tens of thousands of pine nuts in several hundred locations, scattered over many square miles. In winter and spring, they remember the locations well enough to recover most of the food.[31] Humans are not nearly so good at remembering where we put items such as keys, gloves, or cell phones that we use every single day despite the small number of items that we typically bring around with us. Such feats of cognitive virtuosity can be accomplished with the relatively small brains of birds, and even insects such as digger wasps not only remember where their larvae are buried in the ground but bring the correct amount of food depending upon how big the young happen to be.

Biologically, humans are products of the same process of evolutionary change as all other species. The notion of inherent human uniqueness or superiority is dated. One by one, the supposed cognitive markers of our special status in evolution have fallen. Other species learn socially, solve mental problems, use elaborate vocal communication, make tools, are self-aware, express themselves creatively, and so forth. Human languages have a semantic structure that may be absent from the communications of other species, although this conclusion is premature. Complex human language did not exist two million years ago, of course, and may have only been around for a mere fifty thousand years.

Whether experimental apes mastered even the rudiments of American Sign Language was controversial, yet chimpanzees use gestural communication in the wild. This natural "sign language" shares many features of human nonverbal communication, such as holding out a hand, palm upward, to beg or turning the cold shoulder when upset or encircling another's shoulders with an arm to communicate solidarity. Whether such gestures should be considered a true language is not clear. In particular, it is hard to determine if the movements have meanings that are separable from the action itself. Is the cupped hand used by a chimpanzee to beg for food the symbolic equivalent of the same gesture used by a human beggar? Or is it simply the stretching out of a hand to grasp any food that might be forthcoming? Such questions are inherently difficult to answer if only because we have no way of interviewing a chimpanzee about its communicative intentions.

While language is commonly floated as one bulwark of human uniqueness, this may be mistaken, founded as it is on a relative lack of understanding of the complex communicative systems of other species. A related claim is the argument that by virtue of our individual experiences of existing in human societies, we acquire a set of symbolic meanings that are peculiar both to our diverse social experiences and to our species. This perspective is forcefully presented by Eva Jablonka and Marion Lamb, who go so far as to claim that there is a "symbolic inheritance system" that is bizarrely one of four basic dimensions of evolution but nonetheless unique to humans.[32] They do not offer any explanation of why evolution would favor this form of inheritance for humans but not for closely related species of primates that are cognitively so similar to humans in most other respects. They do, however, argue that symbolic inheritance is an emergent property of human societies albeit of uncertain ancestry and emergence. On its face, this would appear to be just one more criterion of human uniqueness that is confabulated when all others fail.

Why do so many scientists and scholars have this blind faith that we are somehow superior to all other life forms? This is hard to know except that it has been promoted by theologians for at least fifteen centuries and repolished by leading philosophers like René Descartes. Suffice it to say, there are no unbiased empirical tests that have been used to rule humans in and rule all other species out, and it seems improbable that such objective tests will ever be devised. If so, the claims are not falsifiable, and therefore not scientific.

What do Jablonka and Lamb mean by symbolic inheritance? They provide one example of the pictorial depictions of Christ, newly crucified in the arms of his grieving mother. They argue that his image has symbolic connotations for Catholics that do not resonate with, say, Buddhists or Sikhs. Drawing on such instances, they "choose the use of symbols as a diagnostic trait of human beings because rationality, linguistic ability, artistic ability, and religiosity are all facets of symbolic thought and communication." These authors seem blithely unaware that each of these facets has been demonstrated in nonhuman animals with the sole exception of religion. Religion has likely not been around for more than sixty thousand years based on the ancestry of religious figurines. Does this mean that pre-religious people failed the diagnostic test for being human?

The gist of their argument is that human symbolic inheritance is distinct on account of the interrelationships that symbols have to each other. These systems of symbols "allow people to share a fiction, to share an imagined reality, which may have very little to do with their immediate experiences. This is true of stories, of pictures, of rituals, of dances and pantomime, of music, indeed of any type of symbolic system we may think about. All symbolic systems enable the construction of a shared imagined reality."

How about forager bees constructing the "shared imagined reality" of distant food sources by representing their direction and size for observers of the waggle dance? Or how about wild geese constructing the shared reality of their hereditary migration route by referring to a symbolic memory map? What about songbirds communicating the place of their birth via their song dialect that makes them particularly attractive to local females, possibly because they possess local knowledge valuable for foraging or evading predators? Such possibilities are clearly foreclosed by arbitrarily defining symbolic transmission as something that is peculiar to humans, hence making the argument in favor of human uniqueness quite circular.

While humans are the only species with written symbolic systems, writing emerged only in the past five thousand years in early hydraulic civilizations. So the argument that these systems are diagnostic of humanity must be taken with a grain of salt. What about the more fundamental argument that symbols themselves are peculiar to humans? That is also spurious given what we know about animal communication systems, such as the capacity of whales to represent their individual identity in a series of auditory signals that composes their unique signature, that is, a symbolic representation of their individual identity conceptually no different from a written signature.

No other species has an organized religion, of course, but there is no compelling evidence that early humans followed religions either based on the absence of religious artifacts. Proto-religious sentiments are reported in other species, however. Elephants mourn for their dead and even bury them. Attraction to mind-altering drugs is found for raccoons that ingest alcohol in fruit that ferments in the summer heat or chimpanzees smoking marijuana in the company of their trainers in a more laid-back period of academic history.[33]

Claiming that humans are capable of some cognitive feat that other species cannot accomplish typically reflects ignorance about what other species can actually do. It was long assumed that other species were incapable of artistic expression, for instance. Yet chimpanzees developed an unsuspected passion for painting. Their efforts passed a key "smell test" when art experts could not tell the difference between primate paintings and the work of human abstract expressionists. The fact that elephants failed the smell test hardly matters to the broad conclusion. Jablonka and Lamb notwithstanding, we are all products of the same evolutionary process and our inherent commonalities outweigh our differences. For that reason alone, the project of identifying human abilities that surpass everything else in the animal kingdom is vulnerable to contradictory evidence, at least if it is couched in objective empirical terms.

The importance of brain size is easily exaggerated. Encephalization, or the ratio of brain size to body size, is distorted by bodily changes. Whales have much larger brains than humans, but their encephalization is lower because they live in water, a medium that permits the evolution of very large bodies. Humans are more encephalized than other primates in part because they have smaller intestines due to a more refined diet so that indigenous humans have small bellies compared to orangutans, say. The human brain also burns a great deal of energy. These phenomena are quite recent, however, being associated with the use of fire to cook meat that likely arose well within the past two million years. [34]

Magpies and crows have small brains because these are limited by requirements for flight, both in terms of brain weight and in terms of the energy stores that are necessary to keep brains running. Yet corvids manage to do a lot with very little brain tissue and are as good as apes at solving some cognitive problems. (One explanation for this seeming contradiction is that bird brains are miniaturized by having smaller brain cells but more of them, which is analogous to a cell phone having more memory than a desktop computer of the past.) Another shocker is the conclusion that magpies are self-aware, putting them at the very pinnacle of animal cognition.

We are all branches of the evolutionary tree, and it is better to think about human specializations from that perspective than to get hung up on the medieval notion of a great chain of being with humans constituting the link above all the animals. That persistent fallacy animates the human uniqueness debate, where "unique" really signifies superior, or higher up on the great chain of being. Humans of two million years ago were not so startlingly different from other great apes. Of course, a lot has happened to human beings since then in respect to human migration to new habitats and our mastery of many different ways of making a living, from foraging to agriculture to modern global economies. While we have experienced great biological change during that time, many scholars attribute much of our success as a species to information that is socially transmitted. Over-emphasis on social learning is a mistake if used as the main criterion for human uniqueness. This is the social learning fallacy, the view that humans are unique in relying upon hereditary socially learned information.

THE SOCIAL LEARNING FALLACY

Many animals transmit valuable property to descendants that can be used over several generations. Examples range from the reefs where coral insects make their homes to the nests carved out in trees by red-cockaded woodpeckers. Cultural evolutionists like Alex Mesoudi argue that human beings benefit their descendants by passing on to them not just physical products like houses or hunting weapons but a body of knowledge that is critical for survival.[35] The Inuit build igloos with long entrance tunnels that keep out the wind, for instance, and make warm clothing from caribou hides. Expertise in such skills is of critical importance for survival in the Arctic. For many indigenous peoples, knowledge of plants and animals is crucial. Some plants have valuable medicinal properties. Others contain dangerous toxins and require special processing if they are used for food.

Adaptationism has a long and checkered history in anthropology. Physical anthropologists recognized that climate affected body build and knew that the greater stature of the Zulus compared to the Inuit could be explained by the Inuit need to conserve heat in the Arctic and the Zulu need to stay cool under hot summer conditions. Today, there is increasing appreciation of how the human body becomes exquisitely adapted to tool use through fine motor coordination, handedness, and so forth.

In a less successful application of adaptationism, Marvin Harris tried to explain specific food taboos, such as the Jewish rejection of pork, in terms of the costs of having pigs around whether they compete for human food or give people parasites.[36] This hypothesis was not supported by the evidence, and the whole approach, known as cultural materialism, was abandoned. Most contemporary anthropologists question the value of adaptationism in relation to complex human behaviors such as cooperative hunting, but cultural evolutionists differ.

Among the North American Lamilera, the key to success is the high level of organization of cooperative whale hunts where each individual knows his job and is rewarded accordingly when the whale carcass is divided up. Of course, other species hunt cooperatively as well. Lions use complex socially learned techniques with role specialization where some individuals lie in ambush and others drive a prey animal into the ambush. Killer whales coordinate their swimming movements so as to upend an ice floe upon which a petrified seal seeks shelter, and a leader uses vocal communication to coordinate the attack. So what is distinctive about humans that makes us the only "cultural species" according to the cultural evolutionists?

Joseph Henrich argues that the distinctive human edge comes from a body of useful information about how to prosper in challenging environments so that we know how to detoxify foods, make suitable clothing, and build homes that withstand the elements.[37] He illustrates this case by describing many expeditions that came to grief because the otherwise well-equipped European explorers lacked the kind of specialized knowledge that was critical for survival.

Henrich's examples of abject failure range from the Franklin expedition in search of the Northwest passage to the demise of the Burke-Willis expedition in Australia in 1860. The few survivors of such ill-fated expeditions often lived to tell the tale because they were assisted by indigenous people. Spanish conquistadors in Texas found themselves in that situation in 1528 when many of the Panfilo Narvaez expedition missed the ship home and decided to follow the Gulf coast to Mexico. They were helped by kindly Karankawa hunter-gatherers who showed them how to find food before succumbing to less charitable people who enslaved them.

The history of European explorers is certainly replete with nasty stories of intelligent people failing to adapt to harsh new environments. One exception involves the tenth-century Vikings who settled in Greenland and prospered there for almost five centuries. The Viking settlers began as farmers and responded to increasingly cold conditions by shifting to seal hunting.

For a long time, the end of the Viking colony in Greenland was interpreted as a failure of Europeans to adapt, but anthropologists recently collected a lot of evidence that calls this view into question. One important point is that the Viking colony did not fail immediately. In fact it persisted for almost five hundred years—from the end of the tenth century until the mid-fifteenth century. Analysis of skeletons reveals that colonists remained in excellent health throughout their stay.[38] Abandonment of the colony was orderly, given that items of value were not left behind: it was an orderly withdrawal rather than a failure.

The Viking settlement was enabled by the medieval warm period. Emigrants from Norway, Denmark, and Iceland settled on hundreds of farms along sheltered fjords.[39] The success of the colony is indicated by the construction of dozens of stone churches.

Until recently, scholars (including the cultural determinists) had assumed that climatic reversion to cold conditions in the mid-thirteenth century made it more difficult to grow food, resulting in famine and population collapse (after the fashion of the Jamestown Puritan colony). They were wrong. Now a new picture is emerging of a vital trading outpost that exported seal skins and valuable walrus tusks that were used for ivory carvings, particularly in churches.

With a reversion to colder temperatures, agriculture became progressively more difficult. The colonists adapted to declining harvests by turning their attention to the sea. At the end of their stay, they derived most of their nourishment from seals and fish, according to bone analysis, thus mimicking the subsistence economy of the Inuit despite lack of contact with them.[40]

As to why the colony was abandoned, scholars link this to the collapse of the walrus ivory trade that was supplanted by trade in elephant tusks. Demand for seal skins also evaporated so that regular ship traffic from Norway stopped in the mid-fourteenth century.[41]

Without items such as iron tools and lumber, life on the island became increasingly difficult, as well as being isolated and monotonous. In the end, they couldn't take it anymore and went home voluntarily. So much for the historical context. How can we explain the remarkable success of the colony—although scholars often mistakenly use it as an example of failure—given that so many other such projects came to grief?

The standard explanation—namely that kindly natives lent their expertise to foundering European explorers—does not work in this case. The nearest Inuit settlements were many hundreds of miles away in northern Greenland. The Viking settlers were completely on their own.[42] Initially, they made a living as farmers before adapting to a novel seal-hunting way of life similar to that of the Inuit. The transition likely occurred within the lifespan of an individual. Of course, the Vikings did bring some knowledge about hunting in addition to implements like spears and harpoons.

As the climate cooled, colonists gave up cattle because these large animals are difficult to provision through a long, cold winter and switched to raising sheep.[43] Seal meat became their staple food, accounting for about three-quarters of their diet, and they used seal oil lamps as a source of heat and light in homes that were built from sod. The switch from mainly farmed meat to mainly seal flesh probably occurred within the lifespan of an individual. The primary evidence of this comes from dietary analysis of bones.[44] Around 1250, there was notable decline in temperature. Between 1250 and 1300, the proportion of marine food in the diet rose from less than 10 percent to more than 75 percent, a shift that can only be explained by incorporation of seals into the diet that became a staple food source thereafter, keeping the proportion of seafood above 50 percent for most years until the colony ended.

The Viking colony in Greenland proves that people do not need the accumulated knowledge of many generations to master a completely new ecology. This can be accomplished through intelligent problem solving by individuals contrary to the narrative by Joseph Henrich, Alex Mesoudi, and other cultural determinists. These scholars claim that human adaptation to diverse environments is predicated on a body of specialized local knowledge inherited over many generations. The actual success of the Greenland colony in switching abruptly to an entirely new seal-hunting way of life without relevant inherited information punctures the myth that we are exceptional as a "cultural species" totally dependent upon inherited knowledge for exploiting environmental opportunities.

Such rapid adaptation is not easy, of course, which is why so many expeditions failed. The Viking colony in Greenland is one case of successful adaptation.[45] Unfortunately, the details of their lives are obscure. We do know that they survived and were very healthy on a new diet of seals. Their success reflected what they learned as individuals, as opposed to benefiting from inherited expertise relevant to an Inuit-type subsistence. Whether the knowledge base is of recent or ancient origin, there is no question that human success in any ecology is predicated on unusual levels of cooperation, including shared information that facilitates survival and reproductive success.

OF SHARED KNOWLEDGE AND COOPERATION

Cultural determinists emphasize that members of a community know far more than can be mastered by any one individual and that the volume of information transmitted via human social learning generally accumulates over time. These claims are truisms today at the height of an information revolution involving exponential increases in digital information. They would not have been so obvious in the past.

One important form of knowledge concerns technological advancement and the emergence of a refined tool kit. For most of our history as a species, there was close to zero development in the sophistication of stone tools, such as hand axes. Local improvements did occur, but they neither spread geographically nor were retained across time.[46] The cumulative aspect of human technological change is recent, denting the claim that *Homo* is uniquely "cultural." On the other hand, there is evidence of extensive anatomical and physiological adaptations (via conventional gene selection) for increased manual dexterity following the development of fully bipedal locomotion in *Homo erectus* more than two million years ago.

In addition to the fully opposable thumb that other great apes lack, humans have fine motor control of individual fingers that permitted our ancestors to strike prey accurately with stones and other projectiles. These adaptations facilitated the eventual use of spears and bows and arrows that appeared around fifty to one hundred thousand years ago. So much of our adaptation was driven by natural selection rather than accumulation of socially transmitted knowledge. Of course, much of this evolution was driven by behavioral change, whether it was adjusting to life in the grasslands or holding a rock steady with one hand while striking it with a hammer stone wielded in the other. Behavioral change of this sort does not require social transmission of ideas, much less cumulation of knowledge. All animals change their behavior when they adapt to novel ecologies, and that fact does not make them "cultural."

Cultural evolutionists see culture as information that is out there in the sense of being transmitted through a different channel than either genes or individual learning. This approach suffers from a lack of clarity about what information is—whether it involves behavior, tools, symbolic depictions such as books, or artistic representation, or whether it is the accumulation of socially acquired information we carry around in our heads.

However it is defined, cultural evolutionists see socially learned information as of considerable importance in human societies and the diversity thereof. For example, Mesoudi points to evidence that people from different countries vary in how interested they are in civic duty as assessed by reading newspapers, voting, having high levels of social trust, support for gender equality, and so on.[47] So Swedes are higher on civic duty than Italians are. If civic responsibilities are socially learned, then immigrants to the United States would be expected to score higher on civic responsibility if they hailed from Sweden than from Italy. This hypothesis was supported, and Mesoudi uses this fact as critical evidence supporting the existence—and importance—of culture.

Is it? Not really! The problem is that our information about how civic engagement develops is rather limited. If I had to guess, I would suggest that civic duty is affected by individual experiences in complex ways. For instance, a child who experiences corporal punishment might have a less rosy view of authority and be less interested in voting or have lower social trust. Similarly, being raised in a large family where money and food are scarce may sharpen competitiveness and cramp civic engagement. Then there is the issue of how much economic redistribution there is from wealthy to impoverished members of the community. Is there an effective government safety net? What is the level of economic development? How effective are public schools? Each of these variables probably affects what individuals learn about the desirability of helping others in the community and being conscientious in fulfilling social duties. In the absence of detailed information about each of these developmental factors, and possibly many others, claiming that there is a culture of civic duty that is transmitted exclusively through the social learning channel is mere hand waving. Absent detailed empirical research, referring to socially learned norms of civic duty that determine civic activities in different countries is an exercise in circular reasoning. There are group differences in civic responsibility. We do not know exactly why these differences exist and describing them as "cultural" explains exactly nothing.

As to the specific question of why the civic duty of American immigrants resembles their ancestors, there are many plausible explanations—including genetic similarity—that have little to do with inculcation of beliefs about the desirability of civic participation (that is supposedly measured in attitudinal research). The mere fact that we do not normally think about these other mechanisms does not mean that they are unimportant to civic duty or social behavior more generally.

Brain development and individual learning are two obvious channels through which civic duty gets shaped in early life. Another is epigenetics. Research motivated by the cross-generational cycle of abusive parenting (that may be interpreted as the antithesis of civic duty) indicates that exposure to physically abusive and emotionally neglectful parenting affects the expression of genes in the second generation so that they are more likely to behave in abusive and neglectful ways toward their own offspring.[48] These mechanisms have been reasonably well established in cross-fostering experiments using monkeys. Such experiments originate from extensive research showing that juvenile rats are highly sensitive to the amount of maternal attention they receive. Of course, such experiences are important because they affect the developing brain. Some of these non-social-learning mechanisms persist across multiple generations during which immigrant communities adjust to their new location.

As another test of the importance of "culture," Mesoudi looked at societal differences in responses to the Ultimatum Game that studies how subjects split up free money.[49] In this game, the first player (proposer) makes an offer that the second player (responder) may either accept or reject. If the offer is rejected, neither player gets any money. Among American college students, the most common offer is around 50 percent of the sum and this is considered fair. Yet, in small-scale societies, there is typically less of a sense of fairness. The Machiguenga of Peru and the Hadza of Tanzania typically offer only about a quarter of the money to the other player, keeping around three-quarters for themselves. Acceptance of offers also varied widely, with the Quichua of Ecuador never rejecting any offer and the Au of Papua New Guinea rejecting more than a quarter of all offers.

The offers of the Lamilera, who practice cooperative whale hunting in Indonesia, were as generous as those of people in developed countries. Mesoudi argues that other subsistence societies may be less generous because members are less involved in the monetary economy and less practiced in making deals. On the other hand, the Lamilera may be more generous on account of their participation in the cooperative whale hunt that involves a detailed division of whale meat according to the specific role of the hunter. That may well be true. Yet, in the absence of relevant data, we do not really know whether the Lamilera are unusually generous in other respects. Given that women do not participate in whale hunts, there is no particular reason that they would be unusually generous if tested in the Ultimatum Game. The Ultimatum Game lacks external validity for indigenous societies that rarely use money.

Despite this, Mesoudi has no trouble in leaping to the conclusion that there is a norm of generosity among the Lamilera in general that is not found among other small-scale societies. There are several scientific problems with this approach. To begin with, norms are not available for direct empirical analysis and rely on self-reports at the best of times. They are intervening variables that are inferred as putative causes of behavioral predictability. Cultural determinists are fond of such intervening variables because they are easily assessed using attitude questionnaires.

Note that Mesoudi's use of norms of generosity is inherently circular. He argues that cooperative whale hunting establishes norms of generosity among the Lamilera that affect their responses in the Ultimatum Game. Why not just conclude that Lamilera generosity is an adaptation that facilitates cooperative whale hunting (as it very well might be)? After all, invoking norms adds nothing that we did not already know. The answer is that Mesoudi invokes norms so as to make a rhetorical case that the explanation is "cultural." After all, norms may get passed around from some people to other people. In summary, Lamilera generosity in the Ultimate Game is attributed to an (unmeasured) norm of generosity that is reified (or fabricated from an idea). In other words, generosity explains itself. So Mesoudi's explanation is vacuous and circular: it is not a scientific explanation at all.

Remember that Mesoudi is using the Ultimatum Game data as cast-iron evidence of the importance of culture. In reality, all these data show is that there is diversity in how societies around the globe respond in this one laboratory game that likely has little external validity to begin with. He does not establish that any of these differences are due to socially learned norms because he has not ruled out alternative explanations. Indeed, he has not assessed any of the probable causes of differences in generosity ranging from genes and epigenetic effects of childhood trauma and corporal punishment, to competition among siblings, to parental investment, food uncertainty, initiation rituals, or social inequality, all of which are known to affect generosity.

Mesoudi's third example of the importance of culture relates to different thinking styles in East and West, specifically the fundamental attribution error.[50] This involves a tendency to attribute behavior to personality rather than circumstances. If a person shows up late for work because they got stuck in a traffic jam, we might instead imagine that they were late because they were too lazy to get up on time. Researchers find that the fundamental attribution error is larger in Western societies than in East Asian countries where people pay proper attention to situational influences on behavior. We do not really know why such cognitive differences occur, but calling it "culture" is very far from a satisfactory explanation.

Note that there are many possible individual-level explanations that are not even considered. For instance, there are geographic differences in personality that suggest genetic variation and might plausibly influence willingness to hold other individuals accountable for their actions.[51] East Asian countries also have rather high population density that could have the effect of increasing stress hormones and inhibiting confrontation. This might involve more willingness to consider situational explanations for evident personal failings. Another potentially relevant difference is that children in some East Asian countries master pictographic scripts, an educational experience that is known to alter how the brain works precisely by increasing focus on the overall picture at the expense of fine details. Genes, brain development, and hormones are all plausible individual-level explanations for geographic differences in the fundamental attribution error. With these possibilities—and countless others—remaining open, it is hard to conclude, as Mesoudi does, that the greater fundamental attribution error in the West establishes the importance of culture. All we really know is that there is some geographic variation in the fundamental attribution error. We do not know why it exists and cannot therefore claim that it is due to differences in social learning in different places.

Mesoudi's three examples of the importance of culture fit a common pattern: find a phenomenon whose societal variation is incompletely understood and claim that the differences are due to "culture." Imputing the variation to "culture" has no more scientific credibility than saying that it was put there by God. If we do not understand the relevant underlying mechanisms, we know nothing of any scientific value. Unfortunately, cultural evolutionists fall into the same epistemological trap as other cultural determinists.

That said, it cannot be denied that students of cultural evolution developed computer models that help us to understand changes in phenomena as diverse as arrowheads, business corporations, and empires. This approach involves a generalization of Darwin's theory of evolution to nonbiological phenomena.[52] It begins with the premise that if there are three key constituents, evolution may occur. These are variation, competition, and reproduction, from which Darwin pieced together his theory of evolution by natural selection. In Darwin's formulation, favorable variants (variation) are naturally selected (competition to survive and reproduce) and pass on their beneficial traits to offspring (via reproduction).

One of the most interesting applications of this approach is in archaeology where phylogenetic methods from evolutionary biology were used to reconstruct the history of artifacts such as projectile points used by North Americans from about 11,500 years ago.[53] Such objects undergo descent with modification, much like biologically transmitted traits. Careful statistical analysis revealed that artifacts like arrowheads follow a similar branching pattern over time as biological traits do. This implies that the artifacts are subject to the same process of natural selection as their owners. If this conclusion is correct, it also means that preagricultural technologies were inherited vertically, in other words, from parents to children. This means that there was no diffusion of arrow design across populations.

This conclusion has two important theoretical ramifications. The first is that if stone artifacts descend with modification over time in exactly the same way as biological traits, then an evolutionary analysis is relevant to items we make or construct even if they have no obvious genetic correlates. This means that useful objects are taken along for the ride with successful populations. For instance, ancestral Inuit who began making clothes from the skins of caribou were likely much more successful than those who relied on other materials. In doing so, they were obviously co-opting a biological adaptation by the caribou to cold conditions. By the same token, animal tool traditions and animal constructions, such as weaver bird nests, can be analyzed *as if* they were genetic adaptations whether they have an identifiable genetic basis or not. After all, natural selection is acting on the phenotype directly rather than indirectly through genes.

The second implication is that diffusion of technological innovations across communities or societies must be extremely recent. With diffusion, the vertical pattern of descent with modification gets disrupted as traits spread geographically in a very short period of time. Such "horizontal" transfer of information is illustrated by the rapid diffusion of cell phones around the globe. This means that residents of societies where cell phones are used benefit from a technology that few people understand or could recreate.

Technological diffusion means (a) that a society uses a lot more information than can be mastered by any one individual, which is a major departure from small-scale societies, and (b) that the complexity of technology can increase rapidly over time. It is difficult to place a date on the origin of cumulative technological innovation, but this enhances contemporary human ecological flexibility, which greatly exceeds that of other vertebrates. Of course, the same is true of cumulative knowledge more generally. Human societies are becoming progressively more complex. Why did that happen?

5

THE EMERGENCE OF COMPLEX SOCIETIES

Human adaptation is like a play with many different scenes. Our remote hominid ancestors evolved in southern Africa. Compared to other primates, they changed rapidly in anatomy and physiology, as well as behavior. In becoming adapted to grassland habitats some four million years ago, ape-men (Australopithecus) began to walk upright, as illustrated by the fossil Lucy. Bipedalism offered greater visibility as our ancestors scanned the horizon for predators or for prey. It left the hands free for carrying tools or food. Walking upright also reduced exposure to the sun's rays, and modern humans are adapted for coping with intense heat by being mostly hairless except for the head, which is protected from sunburn by thick hair. We have more extensive sweat glands than any other species. Of course, the purpose of sweat glands is to promote cooling via evaporation.

Since Lucy—who was little more than four feet tall—hominid stature has increased substantially, and there has been more than a doubling of brain size. The size of the intestines shrank considerably, freeing up energy for a brain that consumes around a fifth of the total energy used by the entire body. Better manual dexterity facilitated more precise fabrication and use of tools.

Considered purely as physiological machines, our bodies changed from a fairly run-of-the-mill ape to something completely different within just a few million years. The most interesting part of the story is not our extensive bodily transformation but what we managed to do as modern humans, using our large brains. Human encephalization likely reflects the multitude of problems we encountered as we dealt with the ongoing challenges of new habitats and varied climates.

ADAPTATION AS GEOGRAPHIC DISPERSAL

After colonizing open grasslands (associated with wetter climatic conditions), the second major phase of our geographic transition was moving out of the African homeland. This transition likely began around a million years ago based on archaeological evidence that places *Homo erectus* in Israel eight hundred thousand years ago.[1] *Homo erectus* is probably not the direct ancestor of modern humans (who is thought to be *Homo heidelbergensis*), but if *Homo erectus* made it this far, it is likely that our direct ancestors made it there as well.

The most plausible explanation for the move out of Africa is that our ancestors moved in search of better hunting prospects. Human populations subsequently spread over Europe, Asia, and ultimately colonized Australia and even populated the Americas via the narrow land bridge across the Bering Strait thought to have connected North America with contemporary Russia. With a refined tool kit that facilitated the slaughter of large herbivores from a distance, humans managed to crash the populations of large game animals wherever they went, creating the need for domesticated animals and plants as an alternative food source. This stage of expansion gives us a glimpse into the behavioral capacities of humans as a highly adaptable species that was no longer restricted to a specific range where it had originated but could expand into an unprecedented variety of habitats from sub-Arctic ice and tundra to the high Andes and from perpetual life on the nonfrozen seas of the Indian Ocean—after the fashion of sea gypsies—to life in the deserts of Arabia.

The agricultural revolution brought on a ten-thousand-year period of relative sedentism. This transition had numerous biological and social consequences. The biological effects included adaptations to reduced dietary variation. For instance, cereal farmers in northern Europe evolved paler skin so as to maximize sunlight exposure and increase synthesis of vitamin D in the skin (given that this vital nutrient was low in the diet) as already discussed. The social consequences were also profound: accumulation of wealth as storable grains created hereditary inequalities of wealth and status. This facilitated the rise of rigid caste systems and slavery.

The third stage of geographic transition was the development of a globalized trading system based in urban centers. The end stage of globalization is functional interconnection of markets and supply chains around the world. This more affluent and more complex way of life also has numerous biological and social consequences. For instance, the need to handle a great deal more information under time pressure makes us more intelligent. Rising general affluence also undermines religious belief.[2] Homogeneity increases across countries in terms of diet, clothing, government, entertainment, and so on. This phenomenon is exaggerated by global communication networks, such as Facebook and Twitter, that encourage people to see themselves as members of a global society. When increased homogeneity occurs in response to shared environmental influences, evolutionary biologists refer to it as "evolutionary convergence." I argue that the reduced variability of societies following the Industrial Revolution can be interpreted as evolutionary convergence as well, albeit independent of gene selection.

BODILY ADAPTATIONS TO HUNTING

As humans colonized new habitats, their way of life changed rapidly with broad implications for evolutionary changes in bodies and brains. Some two-and-a-half million years ago, early humans produced intentionally shaped stone tools. These were used to butcher meat (as indicated by cut marks on bones) as well as to cut grasses and to saw wood. By two million years ago, brain lateralization emerged based on evidence that tools were made by right-handers. Evidently brain processing capacity was at a premium, driving hemispheric specialization. Some scholars argue that the brain was working harder due to increased communicative complexity, but this is highly speculative.

By 1.8 million years ago, *Homo erectus* had much larger brains than their ape-man ancestors (Australopithecus) at about 800 cc compared to 500 cc. Skulls found in Asia and the Caucasus indicate that erectus were already on the move in their migration out of Africa. This implies a capacity to adapt to novel environments that helps explain the larger brains. At this time there were a number of other bodily adaptations to hunting. Changes in the shoulders and wrists likely facilitated accurate throwing as our remote ancestors used stone projectiles to bring down prey. Their success in acquiring high-energy foods, such as meat, is indicated by smaller teeth and less powerful jaws that permitted facial shortening. With more energy-dense food, the gut was reduced, freeing up energy for larger brains and more rapid processing of information.

By 1.5 million years ago, more refined stone tools emerged. Blades were worked on both sides to produce thinner edges. Around this time, changes in finger bones permitted a better precision grip, indicating that tool use was important enough to be a selective force driving the emergence of modern hands.

The earliest evidence of the use of fire dates to about 1.5 million years ago. Some three-quarters of a million years ago, the first stone hearths emerged.[3] It is possible that humans had made use of fire opportunistically for much longer (i.e., exploiting naturally occurring wildfires) but did not know how to control it or use it to prepare food close to home. Fish and turtles were part of the diet, implying that people were active in or around bodies of water, possibly using dugout canoes that disintegrate over time and are thus invisible in the fossil record. *Homo erectus* also had a sophisticated knowledge of plants, making for a diverse diet. They were skilled at working stone and extracted large stone slabs from quarries that were transported to home sites to be later worked into tools. It has even been suggested that erectus was capable of navigating open waters in dugout canoes some eight hundred to nine hundred thousand years ago based on the movements of residents of the Indonesian island of Flores.[4]

Fully modern humans—who emerged about two hundred thousand years ago from *Homo heidelbergensis* stock that descended from *Homo erectus*—were characterized by a much more refined tool kit, specifically lighter, more specialized tools, such as spears and arrows, that were designed to kill from a distance. Their ancestry is complex with genetic admixture with several extinct species, including Neanderthals, Denisovans, and unidentified archaic African hominins. Neanderthals were much more robust than *Homo sapiens* and occupied Europe at the same time.

Neanderthals interbred with sapiens. Although it is easy to imagine the stronger Neanderthals abducting female sapiens and raping them, microbial evidence suggests more intimate contact, such as kissing, that would have occurred in the context of a consensual relationship, perhaps even intermarriage.[5] Whatever the means of interbreeding, it seems that the slow-breeding Neanderthals were no match for sapiens and went extinct some forty thousand years ago. Why exactly this happened is unclear, but their refined toolkit likely allowed sapiens to acquire animal food more effectively, which would have been increasingly important as temperatures fell. Neanderthals subsisted on a great variety of foods, such as mushrooms, acorns, greens, and shellfish, and may have been mostly vegetarian in some parts of their range, specifically northern Spain. Their large size and huge jaws suggest that they must have ingested fibrous foods that required a lot of chewing and digesting as is true of modern gorillas, for example, which live entirely on green vegetation and have large protruding stomachs in addition to heavily muscled jaws.

Genetically speaking, modern humans were a mixed bag. and each of their genetic ancestors probably had very different ecological lifestyles. Of course, this further punctures the story of any environment of evolutionary adaptedness going back two million years. The mere fact that the earliest modern humans (*Homo habilis*) practiced hunting and gathering does not necessarily mean that there is a constant multimillion-year selection pressure for hunting. After all, the local ecology varied tremendously with different prey species available, varied proportions of animal food being consumed, and different technologies being deployed. Consequently, Neanderthals were very different from sapiens even though both were hunters who were capable of interbreeding with each other. All we can say is that the use of tools created strong selection pressures for fine manual dexterity and fine motor control for accurate throwing of spears and projectiles.

Unfortunately, we know rather little about the subsistence practices of early humans and how these modified the human body and brain. However, we know considerably more about the impact of ecological variation on human physiology and genetics since the Agricultural Revolution. What we do know indicates that humans are genetically altered in response to even a very brief alteration in the subsistence ecology, such as the ten thousand years of agriculture. Arguably, even the two-hundred-plus years of the Industrial Revolution have wrought extensive biological changes for our species, if not genetic ones.

BODILY ADAPTATIONS

Our remote ancestors likely had dark eyes, whether brown or black. Blue eyes are common in northern Europe and emerged only in the past six thousand years, as discussed above. This phenomenon suggests that genetic evolution can be surprisingly fast.

In contemporary evolutionary theory, organisms are well matched to their way of life thanks to selection of favorable genes. Those carrying less favorable gene variants leave fewer offspring.

When the way of life changes profoundly, as illustrated by an altered diet following the Agricultural Revolution, genetic adaptation gets to work and animals (including humans) become better matched to their altered way of life. This is what really happens, but it clashes with the evolutionary-psychology narrative of humans (uniquely?) failing to adapt quickly to environmental change. Lactose tolerance in dairy regions of the globe and alcohol intolerance in rice-growing regions provide compelling examples of rapid adaptation. So does the match between eye color and cereal farming, as already discussed.

Contrary to the evolutionary-psychology narrative, there are numerous genetic adaptations to the Agricultural Revolution. These include genetic immunity to domesticated animal diseases—such as cow pox—as well as the numerous adaptations to dietary changes.

Humans make such profound changes to their own environment—for example by adopting settled agriculture—that this introduced strong gene selection pressures. As a result, human bodies and brains experienced rapid evolutionary change in the relatively short evolutionary time since we diverged from chimpanzees around five million years ago. Our ape-like remote ancestors, Australopithecus, used stone tools to butcher meat some 3.4 million years ago, probably selecting for enhancement of a precision grip in terms of the anatomy of hand bones.

Early humans (*Homo erectus*) were also distinguished by upright walking that is seen as an adaptation to living in a hot grassland environment as already noted. They had lateralized brains with a preference for using their right hands in tool fabrication. Although their brains were still small compared to ours, lateralization is usually interpreted as a sign that there was increased competition for brain processing capacity, implying that they did more with their brains than Australopithecus had. Numerous other anatomical changes occurred in response to a lifestyle of scavenging and hunting. Extensive food processing meant a more energy-dense diet with less need for chewing and digestion, so there was a marked reduction in the size of teeth and jaws and associated facial shortening. The dietary change permitted a reduction in gut size and consequently freed up metabolic energy that fueled a larger brain that had increased to 800 cubic centimeters by 1.8 million years ago.[6] So rapid adaptation has always been a feature of human evolutionary history and it is not some anomalous product of agriculture.

One finds extensive changes in human biological adaptation long before the advent of modern humans, much less the Agricultural (or digital) Revolution. These changes appear to be much more extensive than anything seen for other apes (who traveled less from their sites of origin and experienced less ecological transition). The big picture is that our ancestors were always adapted well to their subsistence ecology and social circumstances. This contradicts the evolutionary-psychology narrative that we are currently adapted to an archaic environment. The evolution of a precision grip in the service of refining stone tools having fine edges is now well understood in terms of hand anatomy and brain lateralization, for example. We know relatively little about the evolution of complex societies based upon extensive cooperation and trade, however. How and why did human societies acquire a level of complexity that outdoes the social insects?

THE RISE OF COMPLEX COOPERATION

The term "cooperation" is ambiguous. In biology, it refers to individuals getting together to solve shared problems more effectively than they could alone. It is better than the term "altruism," which carries serious ethical baggage. Unfortunately, when we speak of individuals cooperating, it often suggests that they have engaged in a deliberative process during which cooperation is selected as a superior strategy to all the selfish alternatives available. Nothing could be further from the truth.

Like economists, biologists are very good at describing competition and its resolution, whether in the Darwinian struggle to survive or the hard logic of competitive markets. They are weaker at dealing with cooperation, which is too often dismissed as some rare anomaly. Whether biologists can explain it or not, however, cooperation is deeply interwoven into the fabric of life. Our complex bodies arose when individual single-celled animals evolved a capacity to function as multicellular animals—a truly mind-blowing instance of radical Darwinian cooperation that is so far unexplained. Even within our cells, one finds the inclusion of organelles, such as mitochondria, that likely began as independent organisms. Our digestive system provides a microenvironment for hundreds of thousands of species of bacteria that help us to digest food. Such extensive cooperation clearly arose as a product of mechanical selective processes that remain poorly understood. Cooperation is widely practiced among plants and bacteria as well as by animals. One of my favorite plant examples is the phenomenon of kin selection among trees that "recognize" kin and help them to survive and grow.[7]

In multicellular animals, the most remarkable feats of cooperation are found among social insects, who inspired Aristotle in classical Athens and remain an inspiration in our attempts to understand complex modern societies of humans. Social insects are notable due to their sterile workers who cannot reproduce themselves. Instead they invest their efforts in the reproductive activity of their sister, the queen. How this level of cooperation got started remains almost as mysterious as the origin of multicellular animals. The mechanism most often cited to explain worker bee "altruism" is kin selection. The idea is that by helping their egg-factory sister—the queen—to reproduce, they are far more successful in transmitting their genes to future generations than if they reproduced independently.

Fascinating as this thesis is, it is incomplete. In nature, it is possible for high levels of cooperation to emerge among completely unrelated individuals in response to environmental conditions. This phenomenon is illustrated by the swarming of locusts, which has many parallels with the aggregation of strangers in modern cities.

THE LOCUST ANALOGY

It is easy to fall into the trap of assuming that members of a particular species always behave alike and that this predictability is built into them by shared genes. Animals are also affected by the consistency of the environment they inhabit. This point is brought home when radical changes in the environment completely alter an animal's way of life. For solitary desert grasshoppers, increased population density alters brain function and turns them into highly social animals.

When short-horned desert grasshoppers come into frequent physical contact by feeding on the same bush, they experience a surge in serotonin output that induces them to eat more and makes them more willing to mate, generating a population explosion.[8] The key event that triggers these changes is a freak rainstorm in the desert that turns the vegetation green overnight as plants put out new leaves and bloom. The female grasshopper gorges herself on the new vegetation, finds a mate, and lays her eggs. Population explodes.

The desert soon becomes parched again and the lush new growth shrivels up. Young grasshoppers abandon the solitary territorial ways of their parents. They run together in their hundreds and feed in close proximity so that they are constantly bumping into each other's bulky hind legs. The tactile stimulation increases the production of serotonin. Serotonin induces calmness that minimizes territorial spacing and increases willingness to mate. Before long, they coalesce into the huge swarms that migrate together and ravage crops.

If the brain function and behavior of a simple animal like the desert locust is so radically affected by what other locusts are doing, it is hardly surprising that the same may be true of highly social animals like most mammals. Rats learn complex mazes more rapidly if they are housed socially rather than being caged alone. So it is hardly surprising that the intelligence of humans—who are one of the most social species on the planet—would be boosted by increased stimulation in developed countries.

Growing up in modern urban environments exposes people to more contact with others and requires them to process a lot of information quickly. With universal education and more leisure reading, the volume of information to which people were exposed increased dramatically. The cognitive demands of urban life range from mere navigation in a complex maze of streets to acquiring complex work procedures and adjusting to a more diverse social landscape where not everyone thinks or acts alike. In the digital world, we have ever more contact with other people and process increasing amounts of information at ever faster speeds, whether for social or occupational purposes.

The digital world offers an unlimited diet of information and entertainment that challenges our brains during most of our waking hours as opposed to the far more monotonous existence of preceding generations.[9] The thirst for information also increases. Even the plots of fiction become progressively more complex and the characterization and scene editing in movies get increasingly demanding.

IQ scores shot up with economic development. Scholars concluded that this is not primarily due to storage of large quantities of information but rather to an enhancement of our generalized capacity to solve problems.[10] In other words we are better adapted to an information-rich environment. Small wonder that IQ scores of developed countries increased some thirty points (although this is not obvious because tests get periodically renormed so that the population average still equals 100).

Economic development functions rather like an enrichment program for our brains. The same was true of the impact of rising population density on the nervous systems of short-horned grasshoppers who respond to physical contact with others. For grasshoppers, an abundance of food affects feeding and sexual behavior in remarkable ways, as we have seen.

Curiously, economic development also affects both the feeding system and sexual behavior for humans. This is illustrated by the rise of premarital sexuality and of obesity. We are inclined to interpret these issues in terms of high-level processes of social influence. Yet it is refreshing to think about them as examples of adaptation to environmental change analogous to the grasshopper-to-locust transition.

Locusts make the most of rare precipitation in the desert by breeding quickly so that their behavior is an adaptation to changing weather. Such adaptive change occurs not just in the lifetime of the individual but in a matter of hours. It occurs without any change in the genetic makeup of the population (although the underlying biological flexibility is doubtless genetically selected).

Swarming is what makes a locust a locust. Interestingly, swarming is determined by changes in individuals. The same principle holds widely in nature. Whether it is the efficient movement of ants on a trail (that beats the performance of human vehicular traffic) or the beautiful coordinated movements of fish schools or flocks of starlings in the evening skies, something that looks holistic or centralized is not. Such feats of coordinated movement have neither a conductor nor a score, so to speak. Instead, they are an emergent property of low-level adjustment by constituent individuals.[11]

Among schooling fish, for instance, individual fish modify their movements in accordance with pressure sensations received at their lateral line organs and move in the right direction to remain close to the other

members of the school without colliding with them. Analogous adjustments keep birds in large flocks from colliding in midair so that they trace
out patterns in the air that are as coordinated as fish schools.

It is tempting to think that a flock of starlings in patterned flight in the
evening sky is answering to some centralized authority, some alpha bird
who calls the dance, so to speak. Yet, here is no known centralized
coordination. The pattern of coordinated flight operates from similar principles as schooling fish. It is entirely bottom-up—determined by the responses of individuals to the movements of their immediate neighbors. In
other words, the pattern we see in coordinated flying is an emergent
property of the low-level responses of individuals. Similarly, the tendency of migrating goose families to fly in V formations may reflect the
aerodynamic advantage to following individuals riding in the wake of a
leader.

Remarkable as the story of social coordination of movement is, not all
social phenomena are bottom-up. In the human case, complex societies
require centralized control and coordination to some degree. Such centralization is very recent, however, cropping up only in the past ten thousand years. However it is organized, there is another curious parallel
between urban societies and locust swarms. Both are a response to food
surpluses related to increased availability of water. The difference is that
locusts exploited naturally occurring rain whereas human engineers exerted control over the water in rivers and lakes. All early cities followed
this hydraulic (i.e., water-based) model whereas modern cities are less
dependent upon agriculture and flourish through trade instead.

THE RISE OF COMPLEX SOCIETIES

Modern complex societies differ greatly from the simpler societies that
preceded them. The number of people in a forager community typically
ranges upward of forty individuals, consisting of married couples and
their children in addition to older relatives. By contrast, early hydraulic
cities had a thousand times as many inhabitants (around fifty thousand).
Of course, modern megaregions, such as the greater Tokyo area, may
have a thousand times more residents again (with tens of millions of
residents).

Population size is a relatively crude index of societal complexity.
Hunter-gatherer societies were quite open, in the sense that a person
might easily leave one group where they encountered interpersonal conflict and join another where they felt more comfortable. This mechanism
helped avoid deadly conflicts, particularly those related to sexual jealousy

(given that extramarital relationships were common).[12] In reality, the forager society was more complex than one would imagine based upon analysis of the subsistence group. It encompassed many more people encountered as communities moved around their home range from one temporary settlement to another in search of food.

Many foragers engaged in limited trade. So one might find items like necklaces that had traveled a distance of several hundred miles. Trade was limited due to lack of personal wealth, however. Its functions were also limited considering that hunter-gatherers were economically self-sufficient. They got everything they needed to live from their home range and took nothing outside of it so that their net impact on the local ecology was close to zero.

This point may be illustrated by the reclusive Machiguenga tribe living on marginal jungle high up in the Andes who combine hunting with limited gardening.[13] Given the poverty of the land, they must move often as the gardens become depleted of nutrients. Observers find it difficult to tell the difference between land that was abandoned by the Machiguenga and virgin forest, implying that even horticulturalists were ecologically neutral in the sense that cultivated land was quickly reclaimed by the jungle.

Indigenous people were mostly in balance with their environment—a balance that was occasionally broken by overhunting. This meant that there was little demand for economically useful trade goods. Anthropologist Marshall Sahlins wittily captured the self-sufficiency of foragers by describing their lifestyle as the Zen road to affluence.[14] Their needs were modest but everything they needed to satisfy those needs was available in the environment close to them. So there was little need for distant trade and what little there was often involved inessentials, such as body ornaments.

Sahlins notes that because they were constantly moving around in pursuit of better hunting, foragers had very limited personal wealth. Whatever they owned had to be literally carried on their backs as they moved—a distance of several miles—from one temporary camp to the next. Everything else they needed was taken from the immediate environment, whether food, clothes, or new tools. Given that couples had to carry their dependent children, they were limited to carrying only real necessities, such as hunting tools.

This point was brought home to Sahlins by the reaction of forager communities to novelty gifts brought by anthropologists like himself. These included worthless items such as Mardi Gras beads, tinsel, and small mirrors. When presented with such novel items, indigenous people, including adults, would play with them delightedly for hours.

Given their obvious pleasure in the items, Sahlins was surprised to find everything abandoned on the side of the trail. Much as the foragers enjoyed their gifts, they did not see them as having any earthly use. So foragers generally keep only those items that are essential to their way of life.

Sahlins concluded that foragers already possess everything they require to satisfy their, admittedly basic, needs. In Stone-Age terms, they are "affluent." Although he might be accused of creative semantics, of playing around with the definition of the word "affluent," there are good reasons for believing that the forager quality of life—while far short of that in developed countries—was actually better than that of the subsistence farmers who came after them: they were better nourished and healthier and had a high level of food security. [15]

The forager lifestyle was not easy, and our distant ancestors encountered many challenges, such as large predators, poisonous snakes, biting insects, thorns, infections, and interpersonal violence (much of it motivated by sexual jealousy). [16] Even so, they had an excellent diet that inspires modern nutritionists who want to enhance health. Anyone who survived childhood might reasonably expect to make it to their sixties (based on the demographics of contemporary subsistence societies). This does not seem like much of an accomplishment by contemporary standards, but their lives were lived in very good health. Compared to contemporary populations, there was virtually no heart disease, obesity, diabetes, kidney disease, liver failure, allergies, or any of the many chronic illnesses that plague modern populations from adolescence on and account for the bulk of medical office visits and expenses.

So agriculture brought many health problems ranging from dietary deficiencies (due to overreliance on staples such as corn or rice) to repetitive stress injuries that are manifested in skeletal abnormalities. Hard repetitive work, such as grinding cereals, led to bone deformities severe enough to be visible in skeletal remains. [17] Agriculture ushered in another major stressor, namely status distinctions. Increasing income inequality was highlighted as a major cause of obesity and related diseases such as hypertension, liver disease, kidney disease, and heart disease (known as the metabolic syndrome) in modern life. [18] Why did agriculture bring widening status differences?

AGRICULTURE AND THE RISE OF STATUS

Forager societies mostly lack status distinctions. It is true that some individuals have higher status than others. The best hunters are admired by other members of the group. They are more attractive to women and enjoy more frequent extramarital flings. The shaman is also respected as a person having useful powers for intervening in the spirit world so as to improve success in the hunt. In many societies, shamans can be either male or female. Despite this, men generally have higher status than their wives, and domestic violence against women is common.

Other than these social distinctions, there is very little social hierarchy among foragers, and status differences are more earned than inherited. Their societies are said to be flat in the sense that one person is as good as another. In these subsistence societies, status distinctions are actively resisted, and anyone who becomes too big for their boots is relentlessly mocked and brought down to size.[19]

It is true that there is a headman (or head woman) who may be responsible for important decisions such as determining when to move camp. Nevertheless, their powers and privileges are limited. The headman is a servant leader who spends a great deal of time resolving disputes between other members of the group. The one sign of hierarchy is that he is permitted to take two wives (in the case of otherwise monogamous communities).

Any inequality is social and political rather than being based on differing levels of inherited wealth. Why do hunter-gatherers have no status distinctions based on personal wealth? The answer, quite simply, is that there is no wealth-based distinction because there is no viable store of wealth. What changed that divided societies up into hereditary elites, hereditary freemen, or hereditary slaves as illustrated by Egypt under the Pharaohs, which is so vividly depicted in the biblical story of Joseph (most likely fictitious)? The answer to this question is complex: different factors likely influenced the formation of status hierarchies in different societies.

One key contributor to status differences was the Agricultural Revolution. The fact that people lived in the same place permanently allowed them to acquire real property. One type of property was the agricultural land itself. Neolithic farmers from Europe's Linear Pottery people varied in status according to archaeological research that examined the quality of grave goods buried with them.[20] Farmers living in areas having more fertile land (loess soil) were better nourished and had higher status based on the quality of their grave goods.

Analysis of the element strontium in teeth was used to assess the type of soil on which subjects were raised. Those being brought up on highly fertile loess soil had a lower strontium 87/strontium 86 ratio (these being different isotopes of the same element). Some farmers were buried with stone adzes. The overwhelming majority of individuals buried with adzes (62/63) were found to have been raised in areas having loess soil. This means that wealthier people (buried with adzes) hailed from more fertile lands that they inherited from their parents in this patrilocal society where females moved when they married. This adds up to a picture of hereditary status differences based upon land quality. Interestingly, females in the burials had come from a much greater diversity of land types, suggesting that many of them moved from less fertile to more fertile land, effectively marrying up the status system where the elite inhabited the best land whereas lower-ranked families were pushed into more marginal areas where it was difficult to make a living.

Differences in land quality may be sufficient to maintain a basic status system in farming communities where the haves occupy the best land and the have-nots experience a much more difficult life being forced to work considerably harder to produce sufficient food. For their part, the loess-based elite have the best of everything. They have surplus tools that they can afford to bury ceremonially. They are very well nourished. When they marry, they are more attractive to women and therefore end up in unions with the most attractive women, who are likely to be healthier and more fertile than their plainer competitors.

So the earliest agricultural societies had status distinctions that affected health and longevity, differences that were not present in forager groups. The benefits of fertile soil were passed on from father to son because sons remained on their native farms while daughters moved away to marry. In addition to property inheritance, agricultural societies also fostered inequality by increasing population density.

Agriculture permitted a sharp rise in population. Foragers require about a square mile of living space per person due to pressure on the population of hunted animals. In early agricultural societies, however, it was possible for a square mile of land to support one hundred people, and there are some contemporary agricultural communities in Africa and Asia where fertile land supports up to ten times this number of people. With higher population densities, there is more potential for social conflict, leading to aggression over contested resources. This phenomenon is illustrated by territorial disputes among the Paiute in North America who settled along fertile riverbanks and went to war over the best land. By contrast, the Shoshone foragers who ranged over wide areas in search of food did not wage war because there was little that could be appropriated or defended.

Going to war requires better political organization and a more powerful headman who has sufficient authority to lead others into life-threatening military skirmishes or all-out warfare among villages. Such warfare was studied by Napoleon Chagnon among the Yanomamo in South America.[21] In warlike societies of this kind, men achieve status through their military activities, specifically by having killed an enemy. Otherwise, men achieve power and status via military alliances among villages that are designed to prevent attacks from warlike neighboring villages.

The Yanomamo are somewhat unusual in that the primary purpose of their military conquests is abduction of women as reproductive slaves. This system of conflict is partly due to the fact that these people have a pronounced bias in the birth sex ratio favoring males, and the excess of male babies remains mysterious.

Horticultural societies like the Yanomamo have somewhat greater status distinctions than forager bands do, in part because groups are larger—at up to a thousand persons per village—so that political leaders have more status that is expressed in having multiple wives. Even so, there are no hereditary elites.

Elites emerge only in societies that have some form of wealth that is transmissible from generation to generation, whether land or money. Inherited wealth may also consist of herds of domestic animals, as illustrated by the biblical depiction of the Israelites. Or it might be the financial assets of the modern world that are easily passed from generation to generation, allowing status distinctions to pass with them.

Under severe drought conditions, herds are vulnerable, however, and they can also be wiped out by disease epidemics. When the Israelites fell on hard times and had no food left, they were forced to go to Egypt to borrow wheat, with the effect of enslaving themselves to the Pharaoh. This biblical story of Joseph reads like a novel and likely has minimal historical content. Nevertheless, it does illustrate the great power that the Egyptian dynasty wielded by virtue of their monopoly over storable food, such as wheat, and the role of agriculture in creating social inequality.

This level of inequality could never have arisen in the Amazon basin because the staple foods, including bananas and roots such as manioc, are difficult to store. Bananas are the staple food, and these rot a few days after ripening. Roots and tubers are also subject to decay. None of these foods can be dried enough naturally under tropical conditions to make them last. On the other hand, cereals like wheat, oats, and rye dry naturally as part of the ripening process and can be kept for many years in waterproof granaries that are sealed against rodent pests.

Neolithic farmers of the Linear Ceramic people (named for their pottery designs) had a very basic status system compared to the steep status hierarchy found in most early cities. Many prominent early urban civilizations grew up around publicly controlled irrigation systems. In Iraq, the city of Uruk emerged as the first major city in Sumeria, some six thousand years ago. This region lies between the Tigris and Euphrates rivers where the soil was so good for growing grains and other crops that farmers had settled there permanently. Uruk was organized around a canal system that made the soil even more highly productive. As in the modern period, highly productive agriculture meant that much labor was freed up from agricultural work, permitting the rise of specialized crafts, such as carpentry, building, and metal work.

Of course, there was a political elite who lived off the work of others and did no subsistence or trade work themselves. Archaeologists believe that construction and maintenance of Uruk's extensive irrigation system required a large bureaucracy that was itself regulated by a steep political hierarchy presided over by an absolute ruler. Many of the earliest cities are thought to have been associated with such a public irrigation system, whether in India, China, Egypt, Mexico, or Peru.[22] Given that the primary function of the political system was irrigation and water control more generally, these are referred to as hydraulic cities. So the Aztecs made raised fields in swampland in Mexico and the Incas irrigated their potato fields using mountainside terraces around Lake Titicaca.

Uruk had a well-developed writing system that was likely used by bureaucrats to administer government. Metalworking also emerged as a specialized craft, and copper cylinder seals were widely used. We know relatively little about the daily lives and government of Uruk's inhabitants but assume that its government, commerce, and social life must have had broad similarities to other hydraulic civilizations, specifically India, China, and Egypt, about which much more is known. A steep political hierarchy likely extended from slaves at the bottom to an absolute ruler with godlike authority at the top. It was a period of political despotism that lasted from the earliest cities up to the development of constitutional democracy in the past two centuries.

Social inequality was justified and reinforced by the emergence of religions based upon all-powerful deities. Both phenomena are exemplified by the Sun God religion of the Aztecs in Mexico with signs of frightening rituals, such as human sacrifice to ensure a good harvest. Similar scenarios played out in other early large-scale societies (or civilizations). In Egypt, the Pharaohs aggrandized themselves with awe-inspiring monuments such as the Great Pyramids that served as tombs and were conceptualized as a portal into the afterlife, consistent with their claims of divinity and immunity from death. Of course, the labor of constructing the pyramids was performed mainly by slaves who served engineers and tradesmen and were likely kept in check by soldiers. Such despotism is associated with increased religious power and intensity.

SOCIAL INEQUALITY AND RELIGIOUS INTENSITY

There are many different kinds of religion, but all postulate a spiritual realm that is potentially more powerful than the everyday world. Scholars highlight two key sorts of difference among religions. One is the number of deities that are worshipped. The other relates to the moral authority of a religion—the degree to which it is taken seriously as a guide for everyday life and the level of moralistic fear inspired by the deity.

Both dimensions of religion are affected by the increasing scale of societies following settlement in permanent villages. Earlier religions typically included many deities that had limited powers and limited spheres of influence, such as control over game animals or influence on fertility. In monotheistic societies, these minor gods were replaced by a single all-powerful deity. With agricultural settlements, high moral gods emerge who are seen as being very powerful and capable of enforcing moral behavior possibly by being all-knowing and all-seeing. At the same time, religious rituals become more elaborate and more costly, in addition to being more centralized under the control of politically powerful individuals and elites.

The size, expense, and regularity of religious ceremonies increased as societies became more politically complex according to both ethnographic evidence and the archaeological record of monuments and churches. This reached an extreme in state-level societies where priests were supported by the government. Religious monuments became more elaborate, culminating in vast public works projects such as the Great Pyramids in Egypt. The Pyramids communicated—and were an expression of—the power of emperors to recruit a vast labor force whether paid, enslaved, or both.[23] The same was likely true of other impressive religious buildings, such as temples, altars, and calendrical structures such as Stonehenge that linked the ruling power with the turning of the seasons. The central idea was that the monarch, as deity, played a vital role in natural events such as the cycle of the seasons and was therefore indispensable for a good harvest.

Centralized religions with their permanent specialized clerical workforce, support staff, buildings, and ceremonial architecture and elaborate monuments were extremely expensive and must have used up a significant fraction of economic production—perhaps as much as a tenth. Indeed, the cost appears much greater than the benefits received by hardworking members of the religious community. Why would the public go along with the expense of grandiose monuments or maintaining a nonproductive army of clerics who likely lived under the luxurious conditions enjoyed by elites?

The simplest explanation is that ordinary people were coerced to obey the dictates of absolute rulers who used both religious structures and religious ceremonies to promote their own aggrandizement and power. Such absolutism could not arise in small-scale societies and was predicated on a monopoly of storable wealth such as grain or herds. Another explanation delves into the psychology of commitment.

This psychological explanation for compliance links the costs paid by members of religious communities to their commitment to the religion. This seeming paradox was illuminated by research on communes that were founded by Utopian communities in the United States. Some asked their members to give up a lot, whether their possessions, contact with the outside world, or sexual expression. Communes that asked the most of their members (in terms of the number of commitments) actually survived longer than those with lesser requirements.[24] So members who sacrifice more are more committed to the commune.[25]

This view is supported by other evidence. Members of religious kibbutzim in Israel cooperated more in experiments compared to nonreligious kibbutzniks. Their cooperative tendencies were related to their degree of participation in religious rituals. Members of religions see their own coreligionists as morally superior to others and attribute the world's problems to members of other religions. Such shared beliefs are an integral part of religious membership and would likely be socially reinforced by frequent attendance at religious services. Islamic fundamentalists still teach the medieval concept that infidels are morally inferior and that devout Muslims have a duty to put all of them to the sword. These bloodthirsty beliefs are included in school texts in fundamentalist countries like Saudi Arabia, at least according to a documentary film shot by dissidents.[26]

Religions are not unique in getting strengthened by ritual costs on their members. The same is true of initiation ceremonies for warriors in tribal societies. The more brutal or painful the initiation rites, the more committed tribesmen are to their communities.[27] Adolescents from the Thonga tribe in southern Africa undergo a particularly brutal initiation that goes on for three months. This involves beatings, exposure to cold, thirst, eating nauseating foods, and the threat of death for noncompliance and culminates in a violent circumcision.

Life in extremely religious countries, such as Iran and Saudi Arabia, is no bed of roses. Religious penalties include amputation of the hands of thieves and stoning to death for adultery. Yet these countries are characterized by high levels of religious commitment whether as a matter of belief, fear, or both. Both countries have steep social hierarchies with different kinds of elites. Saudi Arabia is a religious monarchy, but Iran is a theocracy ruled over by the ayatollahs under the direction of the supreme leader (Ayatollah Khamenei). Iran used to have a monarchy that was overthrown by the Islamic Revolution. In each case, though, religion is allied with the government and supports the existing power structure.

Religious beliefs often provide a justification for the arbitrary assumption and use of power by secular authorities, from the divine right of English kings to the deification of Aztec emperors as the guarantors of good harvests. Ironically, some religious doctrines, specifically Christian teachings, are opposed to inequality. For example, the Sermon on the Mount cheers on the downfall of the wealthy and powerful. The Christian religious authorities nevertheless cozy up to the rich and powerful and side with the elite in preserving social inequality. The "sell all thou hast and give to the poor" message of Catholicism is belied by the great wealth hoarded in the Vatican. Religious hierarchies support inequality because doing so bolsters their own status as protected members of the elite.

This paradox was appreciated by Karl Marx, who recognized that religious authorities stood in the way of the revolutionary changes through which he sought to make the Sermon on the Mount a practical reality and put the workers in charge of their own government. Marx perceived that religion helped the populace to remain calm despite the manifest inequality of the hereditary class system that he wished to overthrow. Religion provided a justification for inequality. It sapped the revolutionary fervor of the masses, rendering them passive in the face of injustice and politically inactive. It was the opium of the people.

Marx recognized that religion justified the hereditary class system. Just as monarchs drew legitimacy directly from God, the class system was also interpreted as the product of divine providence. Religious adherents were encouraged to obey all their lawful superiors, which included not just the monarch and government but also the aristocracy who enjoyed hereditary wealth and political power. Under the English feudal system, for example, the local squire inherited the right to administer local justice. This fact is expressed in the satirical couplet:

God bless the squire and all his relations
And keep us in our proper stations.

Whatever religious dogmas might say, organized religions have always endorsed the ruling authorities. Doing so allowed them to remain in business.

So it appears that as societies became more unequal following the Agricultural Revolution, religions became more intense, and deities were perceived as more powerful and moral. The all-knowing deity sees everything an individual does and weighs each of their actions in the scale of morality, ultimately threatening them with everlasting torture in hell (according to the Judeo-Christian tradition). In addition to being the opium of the people, religion served as a virtual police force holding the masses accountable for rebellious thoughts and deeds. So religion is forever changing as an adaptive response to varied ecological conditions, not least in its atrophy within developed countries.

Adaptation clearly did not stop with the Agricultural Revolution. We see this in the evolution of pale-skinned European farmers and in the distribution of lactose tolerance in dairying regions and alcohol intolerance among rice farmers, to mention a few of the many striking examples of physiological adaptations to agriculture. In addition to bodily adaptations to agriculture, we see that the form of religion also changed to accommodate a more unequal social system. Post-agriculture religion buttresses a highly unequal and manifestly unfair social system, even when claiming to do the opposite (specifically in Christianity). So the social characteristics of a society are partly an adaptation to the subsistence ecology.

Most evolutionists consider that Darwinian evolution stopped dead in its tracks long ago because most of the people being born survive to maturity thanks to the achievements of scientific medicine, thereby taking out any possible selection pressure. Even if this argument were correct—and it is not—there is evidence of adaptive modification in human populations not just with agriculture but even following the Industrial Revolution just two centuries ago.

ECONOMIC DEVELOPMENT AS EVOLUTIONARY CONVERGENCE

People living in a developed country look and behave very differently from their agricultural ancestors. Clearly, these differences are not due to genetic changes. Yet they may be adaptive: intelligence increases in response to greater information processing requirements of modern life, for instance.

Students of the social sciences are offered two rather implausible alternatives for explaining human adaptation to the modern environment. Evolutionary psychologists believe that human behavior is adaptive but that we are adapted to a way of life that our ancestors led from two million years ago instead of to modern conditions. Cultural determinists believe that our behavior is entirely homocentric: it is divorced from the natural world making adaptation irrelevant.

Both of these theories are unhelpful if we want to understand how modern populations are changing to deal with the challenges of surviving and reproducing. Of the two, evolutionary psychology is the more credible because it is, after all, a scientific theory. Cultural determinism is not, as I have been at pains to show. The key mistake that evolutionary psychologists made was to place genetic determinism at the center of their explanatory framework and to ignore other mechanisms of change. Yet the adaptive changes we see in modern life are independent of gene selection and evolutionary psychologists are at a loss to explain them. Given what we now know about the capacity of other animals to respond to changing environments within a generation or two, it is incongruous to claim that human beings are alone in their inability to change nimbly in response to environmental change. The truth is entirely different. Humans master new environments with a facility unmatched by any other vertebrate, even to the extent of potentially colonizing other planets if contemporary space pioneers are to be believed.

By pinning their theory on genetic determinism alone, evolutionary psychologists pitched a theory that is irrelevant to modern life, unless you accept that complex human behavior is genetically determined and plays out similarly in all generations due to genetic heritage. That approach has serious scientific shortcomings at all steps of the argument. There are no genetic programs that determine complex human behavior, and most of the things we do are reconstructed in every generation based on our experiences.

Of course, the same is true of other species. Local populations of chimpanzees fish for termites in completely different ways, using different tools. It would be silly to claim that fishing for termites is genetically encoded in the face of this evidence. This foraging activity is clearly an emergent property of various chimpanzee attributes that *do* have heritable genetic components, including opposable thumbs, manual dexterity, large brains, intelligence, sociability, and so on. Being tied to genetic determinism is a real disservice if one wishes to understand how and why chimpanzees fish for termites, however. Yet fishing for termites is obviously an adaptive behavior in the sense that it helps chimps to survive and reproduce.

Chimpanzee fishing is adaptive because it solves a problem and helps them to survive by providing a valuable source of protein-rich food. Although chimpanzees solve this problem in different ways in various local communities, they are nevertheless solving the same problem— getting termites out of the nest and into the chimp's stomach. It really doesn't matter *how* the problem is solved. What matters is that it does get solved. Despite the variations in how chimpanzees fish for termites, there is a striking and unmistakable similarity in what they are doing. Termites that had been minding their own business deep in the nest suddenly find themselves drowning in chimpanzee gastric acid. That common pattern is what is distinctive. Solution of the same problem is what defines fishing as adaptive behavior.

We can see patterns across human societies analogous to chimpanzees fishing for termites in the sense that problems of adaptation are solved. One country is superficially very different from another. Yet there is a striking underlying pattern across all modern societies as they pass through industrialization and undergo modernization with strikingly similar behavioral responses. Some of the most striking of the changes that accompany economic development include reduced fertility, declining religiosity, and the rise of gender equality.

Why are there so many similarities between individuals in developing societies around the globe? The simple answer is that all are behaving alike because they encounter similar environmental conditions that pose similar adaptive problems. Family size declines because the cost of raising children in developed countries is so high despite the relative cheapness of some necessities, such as food, in terms of the number of hours of work needed to pay for meals.[28] Such convergence of initially very different societies to similar patterns of behavior is a case of adaptive change even if it occurs without gene selection.

Convergent evolution occurs in nature when the environment drives similarities among unrelated species. Dolphins and whales look very like fish, for example, even though they are warm-blooded mammals and have to come up frequently to breathe air with their lungs, as opposed to fish that stay below the surface permanently and take oxygen directly from the water through their gills. These marine mammals are very different from fish in many other aspects of their anatomy and physiology. They give birth to live young, and the females nourish them with milk secreted from mammary glands. Their social lives are complex, and they have sophisticated communication systems that remain poorly understood—all supported by a large mammalian brain. They also have bone structures, including ribs, spines, and even four vestigial limbs, that are unmistakably mammalian. Despite being mammals, dolphins look just like fish and have a mastery of fish-like movement in water. Why those similarities? The answer is that in adapting to a marine environment, they evolved movement solutions to an aquatic lifestyle so uncannily similar to the adaptations of fish that it can be difficult to tell them apart.

Convergence is quite common throughout Darwinian evolution. One reason for convergence is that there is a limited set of solutions to common problems. For animals living in water, movement is rendered more difficult by the mechanical drag exerted by the medium. Obvious solutions include a streamlined body shape that minimizes resistance from the water and a skin type, or scale structure, that helps an animal to slice through the native element. Students of Australian fossils were startled to discover a species that looked remarkably like a large lion. This was startling because all Australian land animals were marsupials—relatives of kangaroos and wallabies. This marsupial was a top predator and had most of the same adaptations of teeth, claws, and phenomenal strength that made the king of the beasts so successful in Africa.[29]

Animals converge on similar adaptive solutions when they become adapted to similar environments: that is what adaptive convergence means. In the past, biologists focused on anatomy and physiology as solutions to the problems of survival and reproduction, but behavioral solutions can be just as important: they have the distinct advantage of being a great deal faster. Whether it is Israeli black rats mastering the intricacy of eating the seeds of Jerusalem pines in an energy-efficient manner or the Mauritius kestrel shifting to cliffs as a safer nesting site, rapid behavioral changes helped these species remain viable.

As far as humans are concerned, there were two major behavioral transitions—shifting from hunting and gathering to settled agriculture and moving from the Malthusian world of agriculture to the growth economy following the Industrial Revolution. The second transition brought an unprecedented improvement in the quality of life for ordinary people such that contemporary workers live a better life than even the royalty of former centuries. We are better nourished with a greater choice of foods from around the world. Moreover, we are better protected from infectious diseases and have greater life expectancy. We enjoy greater freedom of religion and can publish our thoughts to the world in the twinkling of an eye. Science writer Matt Ridley analyzes the improving quality of life in terms of how much work is needed to pay for illuminating one's home:

> Ask how much artificial light you can earn with an hour of work at the average wage. The amount has increased from twenty-four lumen-hours in 1750 BC (sesame oil lamp) to 186 in 1800 (tallow candle) to 4,400 in 1880 (kerosene lamp) to 531,000 in 1950 (incandescent light bulb) to 8.4 million lumen-hours today (compact fluorescent bulb). Put it another way, an hour of work today earns you 300 days' worth of reading light; an hour of work in 1800 earned you ten minutes of reading light.[30]

Of course, light-emitting diodes are more efficient than compact fluorescent bulbs so that an hour of work today may purchase even more days of reading light. Modern lights are also easier to switch on, safer, and cleaner.

In addition to being objectively wealthier, contemporary people have a much better quality of life thanks to information technology. A time traveler going back fifty years would be none too happy. There would be no smart phones, no Google, no Facebook, no Netflix, no Amazon, no Craig's List, no eBay, no Snapchat, and no Uber. Life would suck!

Agriculture sustained a huge increase in population compared to hunting and gathering thanks to increased food production. Consequently, our species came to dominate the planet in a way that no other vertebrate ever did. Storage of food opened the door to inequalities in power and inherited wealth. It facilitated the rise of hereditary political elites who used money, monopoly, myth, and military power to assert tyrannical control over those who were less favored by wealth and nobility.

This cruel world of hereditary privilege crumbled after the Industrial Revolution because people were suddenly valued for their skills and dedication more than the prerogatives of noble birth. The rewards of hard work became truly meaningful for the first time in history, and social mobility of two kinds materialized. First, increasing productivity drove up earnings so that ordinary people could aspire to a better standard of living than that experienced by any previous generation of workers. Second, gaps opened up between the wages of ordinary laborers and highly skilled workers. Skilled people earned much more than unskilled workers and had greater prospects of social mobility relative to peers even in an economic environment where rising prosperity lifted all boats. This is even more true in today's information economy where tech-savvy people earn much more than restaurant servers.

Work motivation increases along with the monetary rewards for increased productivity. An increased willingness to work hard is the key to modern productivity, although economists ignored this phenomenon until recently, viewing labor as an input to production rather like raw materials. There are four key explanations for increasing work motivation. Residents of highly productive countries enjoy good health and are well nourished so that they are biologically capable of greater work effort and are also more intelligent (due to a combination of prenatal as well as postnatal nourishment).[31] Survival of more males due to improved medical science also means that men experience greater competition to attract a wife, providing a spur to economic ambition. Moreover, workers are economically independent, having escaped the demotivating embrace of an extended family that stifled initiative and rewarded laziness previously.

Why are all developing countries converging in respect to declining fertility, political freedom, gender equality, quality of life, religiosity, freedom of expression, and so forth? If another species underwent a profound change of this magnitude, in this short a time, we would look to changes in their environment as the probable cause. It is as though it had rained in the desert and solitary grasshoppers had begun to swarm. Locusts provide a striking parallel: their social behavior, reproduction, and general brain biology are predictably altered by rain, just as human populations change with economic development and prosperity.

HOW SOCIAL CONVERGENCE HAPPENS

Biology has an important role to play in understanding how modern human populations adapt to the greatly changed contemporary environment. Yet we are forced to go beyond genes and epigenetics to understand how radically we are changing to fit in with the modern economy. Epigenetics certainly matters in the sense that most of the key early influences on psychological development affect gene expression. So when children are exposed to psychologically stressful rearing environments, they are at risk of reduced intelligence, impulsive violence, delinquency, and precocious sexuality. These characteristics almost certainly involve profound changes in brain structure and function that are the product of altered gene expression. Such early influences mean that modern populations are systematically different from earlier generations. Yet the explanatory power of epigenetics is limited. One cannot account for the Industrial Revolution in terms of how gene expression is altered. Clearly, there is a lot more going on.

If not genes, then what? I have been at pains to dismiss "culture" as a respectable scientific explanation for any of the changes surrounding modernization of economies. Yet *social learning* is real, and its scientific respectability is bolstered by its existence among all social animals having a backbone (i.e., vertebrates).[32] Social learning is illustrated very clearly by generational shifts and the naturalization of immigrant populations, such as the Italians in the small town of Roseto, Pennsylvania, who were studied by health researchers intrigued by their low levels of heart disease.

The younger generations of Roseto residents began spending more time with friends than family and acquired their interests, goals, preferences, and habits from peers.[33] These included large houses and outsize meals. These behavioral shifts happened partly through mere exposure to a different way of life and partly due to an exaggerated desire to be accepted by peers, particularly in the case of minorities who seem different from the mainstream.

Another common example of social learning is adoption of cigarette smoking. This is not a passive process but involves considerable give and take. When Americans became aware of the link between cigarettes and lung cancer in the late 1960s, the smoking rate begin to decline and has sunk steadily ever since, illustrating the adaptive nature of social learning (in the sense of enhancing health or survival).[34] In principle, learning to smoke and then learning to avoid smoking is not so very different from small mammals learning what to eat by observing their mothers and learning to avoid toxic foods that are avoided by the mother.

Social learning is real and important for our species as it is for so many others, but it is just one of the many mechanisms through which we become adapted to the very different world that emerged following the Industrial Revolution. The more demanding cognitive environment of modern life induces children to grow up better equipped to handle a vast amount of information quickly—the basic mechanism underlying the Flynn effect of rising IQ scores. Of course, such environmental enrichment effects are found for other mammals, and the level of fear or stress in the early environment is a key influence on willingness to explore and learn.

When children grow up under stressful conditions characterized by material insecurity, they tend to be less ambitious and forward-looking.[35] They try to seize whatever happiness they can squeeze out of the present. They are more impulsive and violence-prone, traits that characterize pre-industrial societies but decline markedly in societies that experience economic growth.[36] So the modern environment of children is consistently different from what it was in the past. Environmental change produces convergence of modern societies with cognitive ability going up, violence going down, and many other changes, including the increased social tolerance and gender equality of modern societies.[37]

People also respond to what others in the society are doing in adaptive ways. Sexual behavior changes in response to varying marriage markets, for instance. In modern societies, most women are sexually active before marriage, and those who postpone sexual intercourse until after marriage put themselves at a competitive disadvantage because men are attracted to less sexually restrictive women. Ironically, in restrictive societies, it is the sexually liberated women who are at a disadvantage, both because they are at risk of unwanted pregnancy and because their marriage prospects are destroyed, pointing them in the direction of poverty and prostitution.

Although scholars are fond of describing such dynamic processes in terms of changing moral values, cultural determinist explanations are circular and scientifically empty. What is really happening is that sexuality and marriage have market-like characteristics. How a person behaves is very much affected by what the preponderance of other individuals are doing, as Elizabeth Cashdan highlighted in her research on "cads" and "dads" in college dating.[38]

So much for the prevalence of premarital sexuality. What about the rise of gender equality that many scholars attribute to changing value systems also? Values explanations are not true explanations, however, and it is better to look for objective causes underlying the movement toward gender equality in wages and political life. Changing work patterns are one good place to look.

Modern women devote far less time and effort to raising children and far more time to generating earnings.[39] There are many reasons for these changes. One of the most important is the availability of effective contraceptives: women find it much easier to regulate their fertility and delay the birth of children in favor of pursuing careers. Another big factor is the emergence of the service economy and rising demand for female labor. Domestic services that had been "free" in the sense of being performed by women in the home without pay got transferred into the monetary economy, whether as paid services or as mechanical devices that performed domestic work, such as laundry machines and vacuum cleaners, which reduced domestic work performed by women at home.[40]

Although many service-industry jobs were and are poorly paid, the great demand for female labor nevertheless empowered women, giving them a more important economic role than was true of earlier generations. Not all of the service jobs were poorly paid, of course, and modern women are enrolled in higher education at a greater rate than men, implying that they compete more vigorously for better-paid careers than men do.[41] Such improvements in economic power almost inevitably lead to greater political power, which is not to deny that women still have to fight for fair pay and equal treatment in the workplace.

Women no longer expend nearly so much personal effort in raising children: they have fewer babies, begin their families later in life, and rely more upon commercial daycare providers. So they can pour a great deal more effort into businesses and careers than was possible for earlier generations of mothers. With a change in the practical realities of their lives comes a need for greater personal autonomy and greater political freedom. These economic changes inevitably alter relations between the sexes: women enjoy greater autonomy in their sexual relationships along with rising economic independence. Gender equality is just one aspect of modernization—one element of the great convergence that makes all modern societies surprisingly similar. That convergence occurs independently of genetic change. Yet it is a predictable response to changes in the economic landscape that helps people to prosper in the modern world (even if it does not boost their reproductive success).

As an evolved species on this planet, we respond to the differing demands of varied environments as other species do, using similar mechanisms including gene selection, adaptive development, individual learning, and social learning. When environments become increasingly similar in modern countries thanks to economic development and globalization, behavioral and psychological convergence results.

This was true of our various adaptations to an agricultural way of life. It also applies to very rapid changes such as those occurring over the past two centuries in tandem with the emergence of continuously growing trade-based economies.

6

A DARWINIAN TAKE ON THE INDUSTRIAL REVOLUTION

The social sciences have generally been very hostile to Darwinism, and this was not helped by the sociobiology movement of the 1970s that proposed invading that territory, even "cannibalizing" it, to use E. O. Wilson's aggressive term.[1] There are several different reasons that Darwinism was attacked in the social sciences. One is the mistaken view that humans are so far ahead of other species in sophistication that any comparison has to be oversimplified. This argument was clearly rejected in chapter 4 demonstrating that animal cognition is remarkably similar to human cognition and sometimes more effective. In general, the anthropocentric stance is defensive and receives little empirical support. Of course, it is also imbued with religious tradition where humans are depicted as the crowning glory of creation. Such value judgments are incongruous in science, and biologists should feel queasy when informed that humans are the pinnacle of evolution because evolution is not progressive. It is simply the result of an accumulation of random changes that are either favored or suppressed given their suitability for contemporary environments. Another major source of objection to Darwinism involves academic history. Darwinism is associated with social Darwinism, philosopher Herbert Spencer's loose interpretation of Darwinian thought.

THE SOCIAL DARWINISM CONTROVERSY

Spencer believed that the Darwinian conflict rages not just between individuals but also between ethnic groups, in a grisly survival of the fittest.

This colonial-era thinking is widely rejected today as having failed in recent history. It was precisely this shadowy theory of a conflict between races that inspired the Nazi movement and their insane efforts to eliminate all minorities in the name of Germanic racial superiority, an idea that was pure fiction.

So entrenched was this prejudice against Darwinian thinking in the social sciences that when E. O. Wilson published his hotly debated textbook *Sociobiology*, in 1975, there was no controversy about most of the content concerning animals. Critics confined their reading to the final chapter that dealt with humans. As an entomologist and ant expert, Wilson had a rather weak grasp of human psychology and offered up much red meat to opponents by incautious statements unsupported by empirical evidence, including the theory that there is a gene for morality. Such armchair theorizing was not good for the progress of Darwinism in the social sciences. Wilson was attacked as a Nazi, who espoused an extreme right-wing political stance. It mattered little that he was just as liberal as the average academic. To Harvard critics, it was as though they had Josef Goebbels in their midst.

These criticisms were mainly motivated by a dark side of modern academic life, namely leftist propaganda, academic censorship, and political correctness. There was also a rearguard action to defend old ideas of human uniqueness and cultural determinism from the perceived threat of biological determinism. It took a long time for scientists to accept that social learning and biological determinism are just different ways in which phenotypes get shaped and that there is no reason to see nature and nurture as mutually exclusive explanations.

On the other hand, there were real academic reasons that Wilson's work received such scathing attacks. Some of these were articulated already, but the key problem for social scientists concerned the views of Herbert Spencer that were revisited by German Nazis in their delusions about the existence of a Germanic race and its supposed superiority to other ethnic groups. It is possibly unfair to Spencer to blame him for the willful misinterpretation of his ideas by Nazi ideologues like Gerhard Eckert. Nevertheless, Spencer articulated a vision of "races" competing with each other for primacy according to a survival-of-the-fittest model of human competition. Even Charles Darwin subscribed to this thesis in *The Descent of Man*.[2]

Such controversies aside, modern evolutionary psychology has not presented a huge challenge to the social sciences: they mostly continue with their blinkered vision of human uniqueness and cultural determinism. Simple-minded genetic determinism is neither a viable option for social scientists nor something that will ever be embraced by them. Economics is something of an outlier. Economists are actually quite receptive to the "bad Darwinism" advocated by Spencer and his contemporaries that influenced the Nazi movement. Economists are comfortable with a competitive world where there is a struggle to control property and wealth and accept that some individuals are constitutionally more likely to succeed than others. Instead of being worried that their ideas will be perceived as right-wing, even Nazi, economists cultivate an amoral world view such that undergraduates experience a decline in ethical responsibility in the course of an introductory course. [3]

Whether it is classified as bad Darwinism or not, economists are uniquely receptive to a Darwinian paradigm. Instead of competing over survival and reproductive success, individuals (and groups) are locked in conflict over money and economic resources. Some economists believe that the capacity to compete successfully is influenced by genotype and social status, so that the Industrial Revolution in England was attributed to the distribution of acquisitive genes throughout the British population via downward mobility of the wealthy elite. [4] It doesn't take long to realize that this is a genetic determinist just-so story that is falsified by other examples of rapid economic growth, such as the Celtic Tiger in Ireland in recent decades (where the wealthier Protestants had fewer children than poorer Catholics).

Whatever one happens to think about this particular theory, it is clear that all of the social sciences (and even the humanities) have a crying need for some large-scale theory that can integrate them and tie them together as Lopreato and Crippen argued valiantly for sociology. [5] The problem is that genetic determinism could not—and will not—fit the bill. With a suitably enlarged kit of causal mechanisms for shaping phenotypes, can the mission of theoretical integration be resuscitated?

Darwin's theory focused upon inherited differences that passed from parents to children by unknown means. From the 1930s onward, the New Synthesis identified genes as the primary agent by which traits are both inherited and selected. This theoretical breakthrough led to a period of incautious faith in genetic determinism that is slowly evaporating as we acquire a more complete picture of how phenotypes are actually constructed during development.

This is a great deal more flexible than the glacially slow process by which mutants crop up and are naturally selected. Gene selection probably takes millions of years to produce fully bipedal forms such as humans. Yet a goat that is born without upper limbs walks bipedally in a single generation, illustrating remarkable developmental flexibility that helps individuals adapt to their specific circumstances. [6] If we are concerned with the process of generating phenotypes that help the individual to survive and reproduce, then there is no logical reason that we should focus exclusively on a single mechanism, genes, to the exclusion of others. Evolution deals with the match between phenotype and environment and is completely blind to the mechanisms by which phenotypes are produced. It follows that biologists and behavioral scientists must include the influence of all known mechanisms in their investigation of pheno-type-environment matches, or adaptation.

What does all of this have to do with economics? Economic develop-ment has changed the environment that humans adapt to so radically that it has changed humans themselves. We are morphologically, metabolical-ly, and behaviorally a very different species from our hunter-gatherer ancestors. So it is impossible to understand contemporary human adapta-tion to the specifically modern environment without some understanding of economic growth.

Evolutionary scientists do not need to take over economics or to domi-nate it in any way. However, they do need to engage with the effects of economic growth (and even their causes) if they are to assemble a plau-sible picture of modern adaptation. The key issue is that we live in a rapidly growing global economy that changes the landscape to which we must adapt and changes us in the process. Up to now, economics has been highly receptive to the "bad" Darwinism of Herbert Spencer, but the unbridled selfishness of that world view is a caricature of what evolution-ary thinking can offer concerning adaptation to the modern world. To begin with, the phenomenon of global trading cries out for an evolution-ary analysis, if only as an exceptional historical pathway among primates.

THE EVOLUTION OF TRADE

Our great-ape ancestors continue to live mostly in forest habitats, with ground-dwelling gorillas constituting an exception. Our closest relatives, common chimpanzees and bonobos, continue their life in the trees and occupy the same ecological niche as they have done for many millions of years. Why did our ancestors descend from the trees and spread across most of Eurasia, reaching Australia approximately fifty thousand years ago and setting the stage for global travel and trading networks of the future?

Exploitation of new food sources is the most common reason that any species expands its range. Before one hundred thousand years ago, our species had reached Israel and begun to expand across the Mediterranean to the west and across the Middle East, the cradle of urban civilization, on the long trek to Indo-China and the Pacific according to Joseph Henrich, a key source for what follows.[7] Our migrating ancestors exploited a fasci-nating variety of foods ranging from shellfish, fish, birds, and ruminants, to roots, grass seeds, tubers, and honey. Each of these foods is energy dense, particularly compared to the high roughage content of a typical primate diet that contains a lot of leaves and shoots. Exploitation of new foods was likely facilitated by the emergence of more refined tools.

The development of a more sophisticated tool technology may be related to an increase in the size of the brain that likely reflects the use of fire to cook meat. Cooked meat requires less intestine for digestion so that the intestines became smaller. Gut tissue uses energy so that a smaller digestive system created surplus metabolic energy that could be diverted to brain tissue (that uses a lot more energy than other tissues).

Anthropologists believe that *Homo erectus* had both butchered animals with stone choppers and used fire to cook meat for 1.5 million years, but they likely took advantage of naturally occurring bush fires. There was at least one population—at Gesher Benot Ya'Aqov in the north of Israel—where hearths were regularly used from 750,000 years ago, however. A fairly sophisticated lifestyle resembling that of later *Homo sapiens* could thus be achieved with relatively small brains. Humans routinely used fire to prepare food from at least 200,000 years ago, based on the age of hearths in many ancient settlements.

By 1.8 million years ago, *Homo erectus* had much larger brains than their ape-man ancestors (Australopithecus) at about 800 cc compared to 500 cc. Skulls found in Asia and the Caucasus indicate that erectus were already on the move in their migration out of Africa. This implies a capacity to adapt to novel environments that may have been facilitated by increased brain size.

Intelligence and a long history of migration across land in search of new habitat were clearly a feature of erectus, as well as modern humans that descended from them.[8] These characteristics were carried forward in time and contributed to the emergence of early trading relationships among Paleolithic hominid populations from about five hundred thousand years ago.[9]

Early trade involved commodities such as food, animal skins, ornaments, and tools that flowed among neighboring populations. Exchange of goods likely accompanied the exchange of mates, given that most subsistence societies practice exogamy (usually with females leaving home). Marriages often feature gift exchanges.

Accounts of trade prior to agriculture are highly speculative, and the same is true of the history of travel by water that was to prove critical for the efficient movement of goods over long distances. In each case, the evidence is thin. Most boats were made of wood that decomposes, leaving no record. Evidence for trading activity is similarly problematic. When we find objects that are far away from their site of origin, this might mean that they were trade goods. Or it could be that they were slowly moved over hundreds of years by migrating populations.

Some eight thousand years ago, obsidian, a type of dark volcanic glass, was used in making knives and other cutting tools. The stone flaked easily into razor-sharp slivers that could be set in arrowheads and meat choppers. Obsidian was probably moved by boat, and the fact that it was worth transporting over long distances implies that it was highly valued, according to William J. Bernstein.[10] Valuable or not, it was difficult to trade over long distances given the difficulty of transportation. Obsidian is only produced in volcanoes, and its specific source can be determined accurately through a unique chemical signature. Due to its weight and the difficulty of transportation, the amount of obsidian found dropped off sharply as one moved away from the source volcano. Virtually none of the volcanic glass was found more than 250 miles from its origin, suggesting a sharp geographical limit on the extent of trade, even for valuable goods.

Contemporary humans are part of a global economy unlimited by distance, however. That fact very much shapes who we are today. Although trading may date back half a million years, its influence on daily life was modest prior to agriculture. Why did our ancestors intensify their trading? Why have we switched over from mostly subsistence economies to mostly commerce? The answer to this question relates to a huge expansion of international trade. One key factor was the development of ever more sophisticated watercraft in the service of distant trade that facilitated the globalization of economies.

THE EMERGENCE OF A GLOBAL ECONOMY

Some seven thousand years ago, traders moved varied products, such as wine and stone tools, along the Tigris and Euphrates rivers in large round boats that carried up to fourteen tons.[11] Some of the products were bartered for valuable animal skins produced by hunter-gatherers to the northwest. The boats were made of a disposable wooden frame covered with skins and carried their own donkeys. The frames were discarded and the return trip was made using donkeys to carry the valuable skins that would cover new frames in subsequent trips. Each round trip took several months. This gives some idea of the difficulty and hence expense of early trade over long distances.

Angus Maddison developed a methodology for quantifying economic production in various countries over the past two millennia.[12] According to his calculations, productivity per person hovered below $200 per person per year in most agricultural societies and was appreciably lower for hunter-gatherers. [13] Of course, these calculations are based on estimates of locally produced goods that are bartered or sold. Foragers do not need to buy or sell much because they are almost totally self-sufficient, or "sustainable," themselves consuming most of what they produce. Quantifying output exclusively in terms of money (or trade goods) is misleading given that it automatically excludes the economic activities of self-sufficient societies.

Economists create the impression that hunter-gatherers are irrelevant to the economy; a more balanced interpretation is that economics is mostly irrelevant to subsistence societies. This is not because they fail to produce or trade goods and services but because they do so outside of the monetary economy. For instance, wives may provide child-care services in return for valued foods, like meat and honey, and protection from wild animals. Similarly, farmers were much more productive than Maddison's statistics would suggest. However, most of the product was consumed by them and their families and thus undetectable in trading statistics.

The phrase "economic growth" can be misleading from this perspective. Part of the growth of modern economies is real, of course, or we would not be worried about global warming and crashing global ecosystems, but some of it merely reflects the transition of goods and services into the monetary economy that were previously part of the subsistence network. Indeed, the modern service economy is partly predicated on monetizing services provided by women without pay in earlier societies. Children were well taken care of in earlier societies without any money changing hands, for instance but daycare is a substantial expense for modern families.

Following the medieval period (ending around 1500), economic production began rising slowly and then, following the Industrial Revolution (some two centuries ago), at an increasingly rapid pace. The global economy, which was essentially flat for all of human history, may now double in just twenty years (assuming an average growth rate of 3.5 percent).

Development is going on simultaneously in every region of the globe thanks to diffusion of technology and expertise, as well as the globalization of economies. Each country is increasingly integrated into the markets and supply chains of others. Such diffusion of technology and globalization of communications networks and technologies is quite recent.

Just two millennia ago, there was minimal world trade despite healthy commerce along rivers, such as that in the Fertile Crescent. The shift from subsistence to commercial societies boosted economic production numbers and resulted in accelerating growth.

Economists are divided about the causes of contemporary exponential growth, but most see a key role for technological development that increases the output of individual workers and facilitates distant trade. Other possible factors include improved human health and longevity and better nutrition (that boost worker motivation and skill development) and increased availability of capital that increases the scale and efficiency of businesses. Growth was kicked off by improvements in water transportation that expanded from rivers to sea coasts and oceans.

THE DEVELOPMENT OF OCEAN TRANSPORTATION

Distant trade used to be prohibitively difficult, dangerous, and expensive. Ever-present dangers included difficult terrain, hostile indigenous people, and thieves. Overland transportation was so difficult that long-distance traders concentrated on a few highly valued goods such as frankincense and myrrh from modern Iran that were moved by camel trains.[14] This trade dates from about 5,500 years ago. Stone monuments from around 4,500 years ago celebrate incense-trading voyages to modern day Yemen and Somalia. These locations were the primary source of frankincense and myrrh that were the primary trading goods of the era. Both products were used as fragrances to counteract the noxious smells of cramped cities that lacked adequate sanitation. Myrrh oil was applied as a body lotion to neutralize the bodily stench of residents who rarely bathed, apart from the wealthiest citizens. Frankincense gum was burned to perfume the air and was used in large quantities in religious rituals.

The incense trade was a source of wealth all along the caravan routes from the source. Local rulers took hefty tolls, taxes, and fees. Indeed, the trade in incense, together with silk and other luxury goods, may have helped tip the Roman Empire into insolvency. Rome received a staggering total of some ten thousand camel loads of incense per year, and that was fine as long as plunder from conquests flowed in. By the second century, "as the conquests ceased, and Romans became ever more extravagant, those more poetically than economically inclined may be forgiven for concluding that the 'power of the empire evaporated in a haze of incense,'" according to historian William Bernstein.[15]

Well before the end of the first millennium, traders had made the grueling overland trip to China via the Silk Road that was later followed by Marco Polo in the twelfth century. Humans transcended the major impediments to foreign trade only when they mastered distance voyaging in elaborate sailing vessels.[16] The Greeks were early pioneers of distant trade by sea.

In the pre-Christian era, the Greeks had few natural resources but generated immense wealth by trading goods such as olives and wine around the Mediterranean, even venturing into the Atlantic Ocean, rounding Europe, and acquiring grain from the North Sea. Their large, finely crafted ships were unmatched by competitors and used a single sail but were mostly propelled by teams of oarsmen.

By controlling transportation, the Greeks also dictated prices so that they could buy goods at very favorable rates and sell them for much higher prices elsewhere. The wealth, government sophistication, and learning of ancient Athens are legendary and influence us to this day, not least in the invention of democracy (with the electorate then composed exclusively of noble families). Ironically, the Athenians acquired much wealth from the slave trade, which was centered in Alexandria, and ferried slaves around the Mediterranean. (Of course, the same contradiction applied to the United States of America where all men were proclaimed to be free despite the thriving slave trade of which large landowners such as Thomas Jefferson were prime beneficiaries.)

Athens relied almost entirely on trade as a means of subsistence, foreshadowing modern trade-based economies, such as that of Hong Kong, as well as the medieval Italian city states and seventeenth-century Holland. There was not enough arable land to support the population, and most fresh food had to be purchased elsewhere.

Talented as the Greek shipbuilders were, it was left to the Vikings to construct vessels that were fast and sturdy enough to cross large inhospitable oceans, such as the Atlantic, according to William Bernstein. Viking longboats had a sail but were propelled mainly by oars, favoring speed and maneuverability, whether across oceans or along rivers (their customary route of attack on raids for plunder). The longboats were constructed using a system of overlapping timbers that permitted the craft to bend and absorb the force of waves, rather than being broken up with repeated battering from rough Atlantic seas.

Made infamous by their raiding expeditions into Europe, the Norsemen were technologically advanced as illustrated by their capacity to colonize Greenland and keep the colony going for some five hundred years, complete with imported farm animals, and based on lucrative trade in walrus ivory that was used in church carvings throughout Europe. [17]

Norway was the center of a huge trading network extending from Scandinavia to Iceland and Greenland and from Scotland to mainland Europe. Contributions of the Norsemen to ocean navigation are often neglected by historians because their craft designs were likely unknown to builders of the elaborate fully rigged sailing vessels that would gain mastery over the seas in the early days of European empires based on trade, including the Portuguese, the Spanish, the French, the British, and the Dutch. These ships permitted major voyages of discovery, such as the expedition of Portugal's leading mariner, Vasco de Gama, to the spice islands and many others, including the trans-Atlantic voyages of Christopher Columbus.

De Gama was a military leader as well and subjugated the local rulers of the spice islands who had previously answered to Islamic overlords. He also targeted Islamic merchant shipping, in part because of a long history in which the Muslim world had excluded Europeans from the spice trade, and Asian trade routes more generally, in addition to hostility to Moorish invaders of the Iberian peninsula. The Portuguese spice trade was so profitable that the country briefly became the wealthiest in Europe given the vastness of the profits for such a small population (then less than a million).

The Portuguese mastered the use of monsoon winds to drive their craft across the oceans of the world (Atlantic, Indian, and Pacific), and their vessels were large enough to carry provisions for many months, which made such long voyages possible. Their navigational skills were pirated by Sir Francis Drake for the British. Drake held de Gama's pilot hostage for a year so as to pick his brains. Sir Francis attacked foreign ships at will, whether Portuguese, Spanish, or Muslim, and accumulated their treasures in his ships, thereby filling a big hole in Queen Elizabeth's budget.

Drake quickly fell out of favor as the British realized that legitimate international trade could be more valuable than piracy, or "privateering," which was criminalized in Britain and other countries. Such laws made the oceans safer for trade. Historians see the opening up of global trade routes as one of the key ingredients in economic growth.

The Francis Drake route to riches and power came to a speedy end, and the concept of sudden opulence after the Portuguese model had a limited shelf life as well. The spice trade may have been very profitable, but there is only so much pepper or nutmeg that needs to be shipped and the market may be quickly saturated. Powerful nations of the seventeenth century saw the benefits of a globalized economy and founded permanent trading posts in remote lands that ultimately became colonies.

Such foreign expansion is expensive, and it was financed by large trading companies that raised capital by issuing stock and received government trade monopolies that guaranteed their initial success. The British East India Company was the pioneer trading company, and its activities were mirrored by the Dutch East India Company. These were huge corporations that functioned like autonomous countries, even to the extent of having an army to defend their ports and trading colonies.

At this time, the Dutch were one of the leading trading countries in the world, and they used public companies where the risks of a trading voyage could be divided up equally among the investors, perhaps in quarters or eighths. If the ship was lost, the merchant would then lose one-eighth of the value rather than the entire value of ship and cargo. Publicly owned trading companies were an extension of the principles of raising capital and sharing risk. Instead of receiving the profits from trading directly, investors in the British East India Company received dividend payouts that enriched large numbers of British investors until the company eventually became bloated, unprofitable, corrupt, and insolvent.

The historical importance of the large trading companies is not just that they constituted the outward reach of viable empires but also that they opened up the globe to trade on a scale that had never been seen before. Along with the movement of goods, there was also a migration of people, and the phenomenon of emigration for political and economic reasons was born. Even before electronic outsourcing, labor had become globalized as people moved from economically depressed regions to more prosperous ones in search of employment.

The global economy expanded in unprecedented ways as cities grew and industrialized. Growth of the world economy over the past two centuries generated a staggering increase in global trade. Ever larger, more efficient, and more numerous cargo vessels transported goods ever more cheaply around the globe, facilitating international trade and unprecedented economic growth.[18] We see this very clearly in the growth of modern shipping where there is a tight correlation between the rise of global trade, as indexed by increased shipping volume, and increases in global productivity (gross domestic product). It is probably no accident that the Industrial Revolution occurred first in England given that the British then enjoyed unequaled mastery over the seas and the ability to trade in far-flung ports and colonies. English cloth manufacturers could import cotton from India or the Americas and send back the finished product for sale, making huge profits for themselves and for the merchants negotiating these transactions.

THE INDUSTRIAL REVOLUTION

Economists and historians have written hundreds of books about the Industrial Revolution. Yet there is no consensus about why this transition occurred first in England rather than elsewhere or why it began when it did. A similar mystery surrounds the Agricultural Revolution when humans abandoned the leisured life of hunter-gatherers for the heavy labor of agriculture, yielding repetitive stress injuries, shorter stature, nutrition deficiencies, and shorter life expectancy.[19] Both events mark critical changes in the way that people made their living.

In each case, economic production rose sharply, permitting a huge expansion in the global population. The Agricultural Revolution has had a greater impact because it has been going on for longer, marking a hundred-fold population increase compared to a ten-fold increase from industrialization. The pace of change in population due to the Industrial Revolution was about five times as violent (with a tenth of the change in a fiftieth of the time).

Why did population rise so dramatically? The simple answer is that we produced much more food following each of our major ecological transitions. With the dawn of agriculture, much greater population density was possible. Whereas a large hunter-gatherer group of one hundred people required at least one hundred square miles of territory to sustain the prey animals upon which they depended, subsistence agriculture could sustain one hundred persons per square mile or more. This simple calculation shows how agriculture permitted a hundred-fold increase in population.

Clearly, urban life permits a much greater density of population again—as many as one hundred thousand people per square mile, implying much greater food production. Cities still take up a rather small proportion of the land surface, and only half of the human population is urban at this point. All those urban people have to be fed and none of them works in agriculture, so farmers have to produce a lot more food than was possible a few centuries ago. This is enabled by mechanization and biotechnology. Scientific farming techniques increase crop yields and meat production. Modern farming constitutes the mature expression of the Industrial Revolution in agriculture.

The Industrial Revolution proper is generally dated to the use of machine looms in the cloth industry in England around 1860. Such claims are inherently arbitrary and debatable, and many scholars hate using the term "Industrial Revolution" and resort to less-loaded terms such as "development" or "economic growth" or "increased production" that are less binary. Of these alternative labels the most boring one—"increased production"—is also the most useful. If the Agricultural Revolution was defined by a sharp increase in food production, the Industrial Revolution involved a sharp increase in production of cloth, food, and everything else. So we get not just an increase in the number of people on the planet but also a sharp and continuous rise in the standard of living whether assessed by material affluence or an increase in the quality of life involving better health and increased leisure.[20] These changes have had enormous consequences for our success as a species and virtually stopped gene selection in its tracks (partly due to low mortality) in addition to generating many changes in our physiology and behavior.

The English Industrial Revolution did not come out of nowhere, and it was preceded by a century of slowly increasing production as machines powered by water and steam replaced work by humans and draft animals. Once productivity jumped after 1860 in England, it roared higher in a continuous trend, and a similar pattern played out in other growing economies after their own Industrial Revolutions. Why did subsistence shift to affluence? Historians and economists have not delivered any compelling causal explanation. Why are some countries on the affluence track whereas a few others remain in the flat subsistence mode of Malthusian societies?

Failure to explain economic growth may stem from insufficient attention to the human side of the problem, more specifically work motivation. Intensification of production occurs not just when one gets raw materials, power sources, machines, and expanding global markets for goods. The Industrial Revolution is an intensification of production, and this never pans out unless people are prepared to work harder in order to produce more. Conversely, if workers slack off whenever they achieve basic subsistence, after the time-honored fashion of all forager societies and most agricultural ones, economic growth never gets going.

THE SAD HORROR OF WORK ON EARTH

The dawn of agriculture is an interesting case study in human motivation. The big event was settling down to sedentary agriculture. We are inclined to think of such major transitions as knowledge-driven—and therefore progressive—but this perspective may be wrong. Generalizing from surviving hunter-gatherer communities, our forager ancestors were plant experts: they knew everything worth knowing about virtually every plant they encountered, its food potential, medicinal properties, or use as an arrow poison. They also knew about agriculture many centuries before settling down on farms.[21] They not only knew about agriculture but actually practiced it. Hunter-gatherers cultivated rye thirteen thousand years ago at Abu Hureyra on the Euphrates—evidently when the supply of food from wild grasses declined.

When nomadic tribes travel about, they come across patches of edible seeds and harvest them. Before moving on, they might scratch the soil and broadcast some of the seeds before smoothing out the soil. Consequently, there will be more for them to harvest the next time around. Cattle herders, such as the Turkana of Kenya, are also opportunistic farmers: in rare good years, women plant sorghum or millet gardens near the wet-season pastures, but most years are too dry for gardening.[22]

The big riddle about the Agricultural Revolution is why our ancestors transitioned from agriculture as a profitable sideline to farming as a permanent occupation. This was not an easy transition, which would explain why our ancestors toyed with farming as a profitable hobby for thousands of years before finally settling down on permanent plots. They knew a great deal about agriculture before they actually became sedentary farmers.[23] There was no revolution in knowledge whereby some Stone Age genius said to his chums, "Look fellows, if you would only stop wandering about, and devoted yourselves to plowing, planting, and harvesting, you could produce a hundred times as much food!"

The problem with agriculture is that although it produces a lot of food, the food is less appetizing than game. Eating a few staple crops, such as corn, millet, or rice, is not just dull fare; it is nutritionally inadequate. Early farmers suffered from nutritional deficiencies, and their growth was stunted. When contemporary hunter-gatherers settle on farms, many complain of meat hunger, suggesting that they feel nutritionally deprived whether that is objectively true or not.[24] Moreover, the hard labor took a toll on their bodies, yielding joint deformities and arthritis. The Agricultural Revolution proved a huge success for the human species. Yet it was bad for the quality of life of individual cultivators. Farmers worked harder, had worse nutrition, were sicker, suffered more joint pain, and had shorter life expectancy. In terms of quality of life, it was a step backward.

The decline in quality of life helps explain the long delay between the part-time practice of agriculture and the Agricultural Revolution when nomadic proto-farmers settled down to sedentary farming. Colin Rudge argues that no sane person would choose the hard life of a farmer over the more leisurely, happier, and healthier existence of a hunter-gatherer.[25] So why did the transition to full-time agriculture ever occur? Rudge attributes the transition to a complex convergence of factors. The most important of these were rising population and rising sea level.

Prior to agriculture, the human population was very sparse. As is true of any top predator, people needed a large hunting territory so that they did not kill off all of the local prey animals. If that were to happen, the humans themselves would experience a decline in population in a well-known pattern where the population of predators is kept in check by the number of prey available to feed them. This dynamic interaction of predator and prey populations keeps the number of predators down and maintains a low population density throughout the predator's range, as was true of ancestral humans.

So how could human population density have risen? Rudge highlights two possible precipitating factors. One was a rise in sea level that reduced the amount of available land. Coastal populations were forced to move inland in Middle Eastern areas where agriculture originated thereby increasing population density. Second, occasional farming rendered humans less dependent upon their game animals so that population could rise even as the supply of game declined.

Whenever the supply of game fell, people intensified their farming activities. Increased food from farming translated into even higher population density. So humans defied the universal principle that predator populations are held in check by the size of the prey populations. With agriculture as a fallback, humans might continue to hunt even after the prey populations got very depleted.[26] In this way, they eliminated all of the large prey animals that had sustained earlier generations and were forced into obligatory agriculture, now their only reliable source of food. (This phenomenon was not restricted to the Middle East, of course, and humans eliminated all of the large grazing animals everywhere they migrated around the globe from Europe, Asia, and the Americas to Australia in the Pleistocene Overkill, a phenomenon that preceded agriculture by tens of thousands of years.)

Whether this narrative of the events leading up to the Agricultural Revolution is correct or not, it addresses a key motivational puzzle. Why did hunter-gatherers turn their backs on the good life they had enjoyed for hundreds of thousands of years and accept a much tougher life on farms with harder work, longer hours, and worse nutrition and health? Perhaps this was not a free choice: it was forced upon our ancestors by the disappearance of virtually all of their main prey and not just the big animals.

With settled farming, humans worked harder, produced more food, raised larger families, and swelled the global population. The Industrial Revolution was also marked by increased productivity of food and other goods that boosted the global population by a factor of ten. It differs from the Agricultural Revolution in two important ways. First, it is about far more than food production: agriculture accounts for only about one-hundredth of the economic activity in some developed countries. Second, it generates unrelenting growth, so that real wages increase substantially during the life of an individual worker. Developing countries become consistently wealthier, and there is a substantial increase in the quality of life as reflected in reduced infant and child mortality and increased life expectancy.

THE QUALITY OF LIFE

When farm workers were forced off the land by mechanization, they experienced poor living conditions. Charles Dickens frequently alluded to the miseries of the urban underclass and their struggles with poverty, pollution, and disease. Such transitions can be wrenching, but the rising tide of prosperity ultimately lifts most boats.

It is not just that wages rise steadily with development, so that ordinary workers can purchase more products. The quality of these items also improves in subtle but important ways. A computer that is purchased today has much greater speed and memory capacity compared to one purchased a decade ago, for instance. Electronic advances permit many innovations like cell phones and drones, along with Google, Netflix, Uber, and many other applications and services that can improve the quality of life.

For many people, these new technologies have become essential ingredients of a good life. For younger people, the thought of being outside cell phone service is depressing. Modern people spend much more time and money on high-tech amusements, and on luxury goods more generally, and less on necessities like food and shelter. This is possible due to the increasing cheapness of necessities. An English farm laborer in the 1790s would have spent three-quarters of his income on food, compared to about one-seventh today.[27]

Not only does the lower food budget free up a great deal of spending power for other nonessential goods, the food is much cheaper in the currency of hours of work and is better in several ways. First, much of it is already cooked and sold in restaurants for maximum convenience. The farm laborer's wife no longer toils for hours to kindle a fire, draw water, and prepare a meal. Our inexpensive food already has a large amount of service included in the price. Another striking difference in quality concerns the much greater choice available today: a modern supermarket sells many thousands of items from around the globe. Ordinary people today have far more dietary choices than the sybaritic King Louis XIV, who had 498 chefs to cook his meals. And the horn of plenty spews out many more goodies than the pampered monarch could have dreamed of. As Englishman Matt Ridley, writing in 2010, points out:

> In the two hours since I got out of bed, I showered in water heated by North Sea gas, shaved using an American razor running on electricity made from British coal, eaten a slice of bread made from French wheat, spread with New Zealand butter and Spanish marmalade, then brewed a cup of tea using leaves grown in Sri Lanka, dressed myself in clothes of Indian cotton and Australian wool, with shoes of Chinese leather and Malaysian rubber, and read a newspaper made from Finnish wood pulp and Chinese ink.[28]

Economic development may threaten the global environment, but it improves the quality of life otherwise. We can deduce this from the country choices of immigrants. Refugees from war-torn Syria streamed into Germany and other highly developed European countries. No one chooses to live in poor nations such as Kazakhstan or Burundi because they expect the quality of life there to be so much worse. Objective indicators of health, wealth, and freedom back up their choices. These differences are not trivial and are illustrated by the near doubling of life expectancy in the United States during the twentieth century as the country went from an agricultural nation, to being heavily industrialized, to having a highly developed service economy.

Development is not just a matter of acquiring sophisticated technologies like the machine loom in England in the late eighteenth century. Such efficient technologies are meaningless if workers are not motivated to work hard and maximize productivity. The key unanswered question about economic development is why workers suddenly became much more hard-working and cranked out more products per unit of time. Rising productivity generated an unprecedented increase in global wealth. This not only enriched titans of industry but also boosted the quality of life of ordinary people to an extent never before witnessed throughout history. Economists pay too much attention to technology in accounting for this transition but downplay its motivational aspects. Economies thrive when individuals strive. People strive when they anticipate a better future for themselves and their families and work to realize that dream.

THE HUMAN SIDE OF THE INDUSTRIAL REVOLUTION: WORKER PRODUCTIVITY

The Industrial Revolution saw a sharp increase in growth for economies that had grown little previously. The annual growth of per capita gross domestic product (GDP) in Western Europe averaged 0.15 percent between 1700 and 1820, compared to 1.51 percent between 1820 and 1998—a ten-fold increase, according to Angus Maddison.[29] One-and-a-half percent might not seem like a rapid growth rate, but it meant that Western Europe was fourteen *times* wealthier at the end of this period than it had been at the beginning—a phenomenon never before recorded in the history of the world.

Why did earlier societies not grow steadily? Relative lack of growth in historical economies is explainable in Malthusian terms where the central focus is on the food economy. This makes sense given that earlier economies were predominantly agricultural. Any increase in food production fueled higher fertility, and with more mouths to feed, the food surplus evaporated. Economic growth was always temporary, as illustrated by the impact of plague in Europe.

The declining population following the Black Death (i.e., bubonic plague) epidemic in Europe boosted economic activity per person. It was not just that there was more food to go around. The scarcity of labor boosted wages and increased social mobility of workers—an economic growth spurt that disappeared as the population recovered. So economies went up and down like a rocking horse but never got anywhere.

In a Malthusian society, increased fertility eats away at any economic surplus until it is totally depleted. One way that this vicious cycle gets disrupted is through a decline in family size. Europe's Industrial Revolution was preceded by family shrinkage. This occurred prior to effective contraception. It is mainly attributable to a later age of marriage. Even though fewer children were born, a greater proportion of them survived, thanks to improving nutrition and health care, so that the population actually increased.[30]

"Post-Malthusian" countries that go through an Industrial Revolution enter a period of continuous growth where the economy expands steadily thanks to a combination of higher productivity and increasing population size. Real wages increase as workers become more productive so that their spending capacity amplifies the demand for goods and services. Post-Malthusian societies are very different not just in being wealthier but also in having lower fertility, greater participation by women in the paid labor force, later marriages, and other profound social changes including more premarital sexuality and increased single parenthood.[31]

Worker productivity is the linchpin of economic growth and therefore the key to explaining why a society transitions form the growthless experience of a Malthusian society to the steady accumulation of wealth following the Industrial Revolution. Economies grow when worker productivity rises: because more wealth is produced, wages rise and consumption increases in a positive feedback system. Why *do* workers suddenly begin producing more per hour of work?

In part, higher productivity occurs when societies get better at harnessing new power sources to run more efficient machines and when labor is more effectively deployed thanks to specialization and the division of labor. Yet, by themselves, improved mechanization and modern management techniques are not enough to get an economy off the ground. British cloth factories in colonial India yielded miserable returns, and industrialists have similar results with factories located in other less-developed countries. Profits are lower in spite of wages being much cheaper. One reason that workers are less productive is that they are less motivated. Growing economies require a secret sauce and that secret sauce is a motivated workforce. Why do workers suddenly get with the program?

We cannot get in a time machine and go back to interview English workers of the late nineteenth century. The next best method is to compare contemporary countries that range from mostly rural Malthusian societies like Burkina Faso or Haiti, to highly developed modern urban states like Singapore or Denmark. Why are employees in some countries working so much harder and producing more goods and services than their counterparts in other less-developed nations? If we could answer this question, we would know why the Industrial Revolution happened and why countries flipped from stasis to growth.

Historians point to many factors underlying the Industrial Revolution, such as energy sources, transportation infrastructure, manufacturing technology, availability of capital, a skilled labor force, the scientific revolution, and so forth.[32] Yet, none of these factors can explain why the Industrial Revolution occurred when and where it did. Such factors can help explain the slow accumulation of wealth in Europe that preceded the Industrial Revolution, however. This slower growth might have been a necessary precondition for the more rapid growth and increased productivity that followed. Yet the growth acceleration in England after 1860, for example, remains poorly understood.

Economists attribute the breathtaking increase in wealth in the modern world to steadily increasing real wages that fuel consumer spending and government budgets alike. Rising wages are predicated on increasing worker productivity, in other words, increased value produced per hour of work; if employers were not making more money, they could not pass some of it on to their staff. Remarkably, economists cannot explain why modern people work harder and produce more than they produced before the Industrial Revolution, and obvious explanations such as improved technology are a surprisingly small part of the story.

Economists Oded Galor and Omer Moav proposed a genetic theory for why the Industrial Revolution began in England rather than elsewhere, as already discussed.[33] They argue that because the English upper classes produced larger families than their impoverished counterparts, their acquisitive tendencies were genetically propagated throughout the population. This social Darwinist theory suffers from lack of generalizability, however. Wealth increased rapidly in Ireland during the Celtic Tiger period of the 1990s on, for example, yet there is no evidence that the Irish elite had larger families. If anything, they were smaller given that generally wealthier Protestants had smaller families compared to Catholics who were historically poorer and raised more children.

Many historians believe that scientific and technological innovations spike economic growth, but such theories are weak because improved technology does not always increase productivity. This point is illustrated by the rapid diffusion of cell phones to contemporary low-growth countries without noticeable increases in economic growth.[34] Although technological innovation spurred the Industrial Revolution, in cloth manufacturing and other industries, in England, improved technology is necessary but not sufficient for growth to occur.

The inability of technology alone to stimulate economic growth is highlighted by the introduction of English machine looms into India (then a colony) early in the twentieth century. Astonishingly, cloth production per worker did not improve for machine looms compared to mechanical ones. Indian plants hired far more workers, but the weavers spent most of their time unengaged in productive work. According to Geoffrey Clark in *A Farewell to Alms*:

> The Indian Factory Labor commission Report of 1909 is full of testimony by employers regarding conditions in the mills. A substantial fraction of workers were absent on any given day, and those at work were often able to come and go from the mill at their pleasure to eat or to smoke. Other workers would supervise their machines while they were gone, and indeed some manufacturers alleged that the workers organized an informal shift system among themselves. The mill yards would have eating places, barbers, drink shops, and other facilities to serve the workers taking a break. Some mothers allegedly brought their children with them to the mills. Worker's relatives would bring food to them inside the mill during the day. "There was an utter lack of supervision in the Bombay mills." One manager even stated that the typical worker "washes, bathes, washes his clothes, smokes, shaves, sleeps, has his food, and is surrounded as a rule by his relations."[35]

So low work motivation was more important than technology in the lackluster productivity of Indian textile mills. Other explanations for increased productivity of developed countries in terms of government institutions, religion, education, scientific knowledge, capital accumulation, and so forth are also weak, at least when it comes to explaining the timing of the Industrial Revolution in England.[36] Such explanations are often confidently asserted without careful empirical testing, even when proper tests are feasible.

Even if technology alone does not account for the increased productivity of the Industrial Revolution, it is hard to deny that advancing technology gets rid of many jobs, thereby driving up the productivity of remaining workers.[37] Technological development increases the productive potential of individual skilled workers, and this means that parents are more willing to invest in education of their children so that they may compete successfully for highly paid jobs, effectively prioritizing quality rather than quantity of offspring. This preference helps explain why fertility declines in developing countries, even as education improves.

So why *do* people in developed countries work harder and produce more? I investigated four explanations for rapidly increasing productivity in developing countries based on the perspective of Darwinian competition of various kinds. Each of these explanations helped explain why productivity growth has been continuous since the Industrial Revolution. They also helped explain why economic productivity in one country exceeds that in others.

FOUR SPURS FOR HARD WORK

The introduction of machine looms was a flop in India because the workers were not on board with the project of cranking out as much cloth as possible per hour of work. There are many reasons why Indian textile workers might have been less motivated than their British counterparts. To begin with, they may have had minimal commitment to the financial objectives of their colonial overlord employers. That interpretation is supported by their informal shift system that cheated the factory owner out of a full day's work. There was a lack of separation between the world of work and the employee's home life including personal activities such as washing, shaving, and socializing. This apparent lack of discipline likely reflected the persistence of an older way of life. It may have masked a desire to undercut the profit motive of the factory owners. Or it may simply have reflected a lack of concern about time efficiency in a society that was not as time conscious as the British were. Britons had an obsession about accurate clocks, built many church clock towers, and loved punctuality. If workers are not very conscious of the passage of time, they can hardly be expected to care greatly about how much cloth they produce per hour of work.

Moreover, there was a lack of familiarity with complex machinery that resulted in frequent stoppages due to breakages. Indian managers assigned far more workers to each machine because they were unfamiliar with more efficient work practices adopted in English mills. The fact that workers brought family members onto the job site implied that their obligations to extended family took precedence over responsibilities to the employer. In societies where people live in extended three-generation families, as was typical of India at that time, family obligations have a way of undermining worker motivation. Conversely, in developed countries where people live in small nuclear families, work motivation can be higher so that productivity can rise. Economic competition among members of an extended family is just one of four key types of Darwinian competition that helps productivity to rise with the level of development.

1. Collapse of Extended Families

It is difficult to be sure why Indian textile workers were less motivated than their English counterparts, but family structure does matter. Achievement motivation rises when nuclear families replace extended ones.[38] In colonial America, for example, extended families were broken up when people moved west, and descendants of these families emphasized self-reliance and achievement training. A similar pattern is observed in other societies when extended families break up, although there is a chicken-and-egg problem because families that migrate, or rupture, may be more ambitious to begin with.[39]

Having a large extended family is a disincentive to economic striving for numerous reasons. If there are other workers in the household, there may be some diffusion of responsibility.[40] To the extent that income is shared, the worker's own quality of life is not noticeably improved by earning more wages. Sharing income with many relatives discourages workers from accumulating savings.

Early Russian economist Alexander Chayanov proposed that agricultural workers on collective farms (or "peasants") worked only to satisfy their basic needs.[41] Doing any more was counterproductive because other members of the household would dissipate any excess resources so earned, and the same is probably true of extended families. This "consumption-labor-balance principle" would likely break down in nuclear families where the benefits of toil mostly stay with the immediate family. When workers migrate to cities leaving their extended families behind them, they are less vulnerable to parasitism from relatives and freer to become more productive. Of course, urban migrants are also forced to be more self-reliant because they no longer have the support of extended family that acts as a rudimentary social welfare system, and this likely motivates them to save for a rainy day.

With increasing urbanization over time, average household size declines. This helps explain why productivity rises. Average household size of a country measures the degree to which worker earnings are vulnerable to expropriation by other household members. The second reason for increasing productivity as countries develop is better health.

2. Improved Health and Reduced Burden of Infectious Diseases

Populations with relatively poor health and low life expectancy are less productive for two good reasons. First, people who suffer from chronic illnesses, such as malaria or schistosomiasis, have less physical energy and need to rest more. Second, life expectancy is lower, so that people live in the moment rather than accumulating money to be enjoyed later in life. In the language of economics, they "discount the future."

Countries can be more productive if residents suffer less from chronic illnesses such as malaria and tuberculosis. Such diseases decline with economic development thanks to better sanitation, improved drinking water, use of insecticides to control mosquitoes and other disease vectors, public health programs, improved medicines, and so on. Improvements in health have been continuous since the Industrial Revolution with life expectancy doubling in the United States between 1920 and 2000 (from 39 to 77 years). Improved health helps explain why people are both able and willing to build a better future for themselves and their families. The toll taken by ill health can be measured in terms of the average number of life years lost to disability, ill health, and early death (or disease-adjusted life years, DALY). Similarly, being well nourished is a major factor in work effort and productivity.

3. Improved Nutrition

Economic development is characterized by a steady increase in availability of food that has striking consequences for productivity. Better-nourished mothers give birth to larger babies that are healthier and score higher on IQ tests. More intelligent children do better in school, are more ambitious, acquire more human capital, and earn more thereby boosting their native economy.[42] Whereas male and female workers would likely be affected much the same by improved health and nutrition and declining involvement with an extended family, the fourth reason for increased productivity—more intense competition over a spouse—applies mainly to men.

4. Competing for a Spouse or Romantic Partner

Males survive at higher rates than ever before, and this intensifies competition over women. Men appeal to women by establishing high social status or having good earning capacity. This is true even in the modern world where women are economically independent.[43] Women avoid going out with men of lower social status than themselves and steer clear of those who are penniless or unemployed.[44]

Men work harder in societies where accumulating wealth is essential for success in attracting a spouse (or sexual partner). This effort boosts productivity. Conversely, in societies where there are "too many women," premarital sex is common consistent with the masculine preference for intimacy occurring early in a relationship.[45] In that environment, men are less interested in striving for economic success and productivity is lower. These ideas are illustrated by contrasting the low productivity of a sub-Saharan country (where adult females outnumber males) with the higher productivity of Europe or China, places with an excess of males.[46]

The proportion of adult men to adult women rises steadily in developing countries for two different reasons. First, with improving medical care, child and infant mortality decline so that far more people reach adulthood. The beneficiaries are mostly male because far more male than female infants and children die in undeveloped countries. When these mortalities are brought close to zero in developed countries (falling from about 30 percent to less than 1 percent), the ratio of males to females rises.

The second reason that more males survive today is that levels of interpersonal violence are much lower, whether one is talking about death in military aggression or homicides resulting from a dispute between two men. Contemporary deaths in warfare are vanishingly small compared to the quarter-or-so of males who died violently in many subsistence societies of the past. Male-male homicides arising from disputes over women may be as much as one hundred times lower although such estimates are necessarily speculative.[47] Violent deaths of men from all causes have shown a steep and steady drop since the Middle Ages, and this decline is accompanied by rising affluence although the shapes of the curves are different with homicides falling fastest early on and affluence rising fastest in recent history.

Men no longer compete directly over women by a show of physical fighting ability. Rather, they must demonstrate their superiority in economic terms by earning money and rising in social status (something that was almost impossible in feudal societies where everyone had a hereditary place in the society, most as serfs who were partly slaves and partly free). The continuous decline in violent deaths adds up over time with males who are saved from dying as infants or children reaching adulthood and competing over women. Male competition increases cumulatively so that men are motivated to work ever harder to boost their earning capacity. There is no reason to believe that women enhance their sexual attractiveness by earning more: even if men prefer higher salaries to lower ones in a partner, such considerations take second place to physical attractiveness.

What is the evidence that men work harder in a tight marriage market? There is not very much relevant research, but most of it is favorable to this thesis. Consistent with the competition-over-women perspective, men were more artistically productive in societies where there was a scarcity of women.[48] Moreover, the scarcity of females in contemporary China increases the practice and amount of bride price payments, so that men must work harder to save enough money to get married.[49] Research on preindustrial societies also found that boys were trained to work harder in polygamous societies where adult male industriousness yields more wives and higher reproductive success.[50]

Over the past half-century, many more women have joined the paid workforce, but their efforts are not designed to enhance their appeal to men. Rather, they work to support themselves in a world where marriage is delayed and a significant fraction of women never marry.[51] Those who do marry mostly remain in the workforce whether to fulfill personal ambitions or to pay the expenses of raising children in a world where these costs are seeing unprecedented inflation.

In summary, the shrinkage of household size, improved health and nutrition, and anticipation of longer life, together with increased competition by men over women, are four plausible explanations for people working harder in developing countries. Working harder would have boosted productivity and wages so that the increased spending power of workers kept businesses and national economies growing. Each of these four reasons for higher productivity helps explain why modern societies generate more wealth continuously, something that never happened before in history. These explanations were tested in a sixty-two-country comparison of economic productivity. What of the findings?

DARWINIAN COMPETITION: THE NEGLECTED REASON FOR ECONOMIC GROWTH

Economists recognize that the big distinguishing factor between economies that grow rapidly and those that do not is worker productivity, or the value generated by an average worker per unit of time. Increasing productivity in developing countries is poorly understood possibly because too little attention is given to worker motivation. My prediction was that productivity would increase with reduced competition over food, survival, and household resources but with increased competition by men over women.

Explanations for growth that rely on physical conditions in a country often come across as just-so stories. For instance, we are told that the English Industrial Revolution was founded on availability of coal as a cheap energy source. Yet Japan, long the powerhouse economy of the world, had almost no native energy sources. The steam engine looms large in stories about the causes of the Industrial Revolution, along with other innovations in cloth production. Yet, when all of that technology was first brought to bear in India, it had zero impact because Indian workers were not motivated to spend their day productively, much less to maximize output. Why were Indian workers so uninterested in competing to improve their lives after the manner of their counterparts in more productive economies? One likely reason is that their health and life expectancy were not as good: there was more of a Darwinian struggle to survive. If workers are to care deeply about their economic future, they need to (a) anticipate long life and (b) be healthy enough now to expend much work-related effort.

This rationale is highlighted by the gruesome impact of tropical diseases in Africa and around the world. Of these, malaria is one of the most familiar and common, but there are many other health challenges in the tropics such as schistosomiasis, or liver fluke, a parasitic disease that infects livestock in temperate climates but brings down people in the tropics. As David Landes writes in *The Wealth of Nations*:

> Warm African and Asian waters, whether canals, or ponds, or streams, harbor a snail that is home to a worm (schistosome) that reproduces by releasing thousands of minute tailed larvae (*cercariae*) into the water to seek and enter a mammal host body through bites or scratches, or other breaks in the skin. Once comfortably lodged in a vein, the larvae grow into small worms and mate. The females lay thousands of thorned eggs—thorned to prevent the host from dislodging them. These make their way to liver or intestines, tearing tissues as they go. The effect on organs may be imagined: they waste the liver, cause intestinal bleeding, produce carcinogenic lesions, interfere with digestion and elimination. The victim comes down with chills and fever, suffers all manner of aches, is unable to work, and is so vulnerable to other illnesses and parasites that it is often hard to say what is killing him.[52]

Considering the health impact of such tropical diseases, it is not hard to understand why there was little economic growth in the tropics prior to the development of potent medicines for such ailments. I wondered whether health problems in contemporary economies bring down worker productivity.[53] In addition to health, I looked at nutrition, considering that workers must expend physical energy that is better conserved through physical inactivity. Do countries having higher nutritional intake—where the Malthusian struggle over food is relaxed—produce more goods per person per hour? I also wondered whether countries having smaller households would be more productive given that competition over wages is limited to close relatives and spouses instead of a large extended family who soak up any surplus resources. If the marriage market is more competitive given a higher ratio of adult males to females, does this drive higher masculine work effort and productivity? In my study, I analyzed productivity in sixty-one countries for which these data were available (with one, Bangladesh, getting excluded as a statistical outlier). What of the findings?

Each of my four predictions about country differences in productivity was supported. Productivity was greater for countries having a higher proportion of males. This bolsters the conclusion that men (who constitute much of the paid workforce in developing countries) work harder if they face a more competitive marriage market. This effect remained even with female labor participation rates statistically controlled. Productivity was lower in countries where people lived fewer healthy years, implying that chronic illnesses sapped work motivation and productivity. Better-nourished countries were more productive. More specifically, productivity was higher in countries where more food was consumed per person. Finally, productivity was lower in countries with more adults per household, suggesting that workers restrain their efforts if the fruits of their labor are divided among a large extended family.

My results provided some insight into the conditions under which worker productivity rises, thereby fueling economic growth. The sex ratio had a large effect. Productivity increases substantially when men encounter a competitive marriage market. Our evolutionary past selected primarily for reproductive success. By that reasoning, men would be expected to minimize economic activities that are irrelevant to reproductive success. If they enjoy sexual access to women without working hard in a job, such effort may be avoided, keeping productivity down in countries having a larger proportion of women in the population.

This phenomenon may underlie the decline in male college enrollments in the United States even as female enrollments increase, putting more women than men in third-level institutions.[54] The steady decline in the proportion of male to female college students occurred over a historical period when premarital sexuality was on the rise. This is no coincidence. After all, why would men strive to be professionally successful so as to marry an appealing woman years into the future if they may sleep with a variety of attractive women now? The desire to party may earn an easy victory over the desire to study. The implications for male human capital, occupational success, and productivity are clear.

Productivity also increases as household size declines broadly consistent with Chayanov's theory about peasant economics that deals with the low productivity of Soviet collective farms once basic survival needs are met. The phenomenon of productivity going up as household size declines might be interpreted in a fairly boring way: fertility declines with economic development so that average household size is reduced because there are fewer children. Yet that interpretation was ruled out by statistical control of fertility.

Apparently, it is the number of *adults* (potential workers) rather than the number of children that is related to productivity. Having more adult members in the household can drain resources, particularly if some of them are unemployed or retired, so that wage earners are less likely to enjoy the fruit of their own labor. Moreover, they cannot accumulate savings given that any economic surplus gets used up by others. The extended family saps work motivation for much the same reasons that being in an agricultural collective did.

The fact that productivity rises for countries with small households suggests a role for increased human capital. Yet this explanation was ruled out by controlling for years of education. One might imagine that when fertility is low, parents would invest more in individual children whose earning prospects are enhanced by expensive educations, that is, a quality over quantity approach to raising children, and that this would boost productivity. Parental investment in children *does* increase as societies develop, but the causal arrow runs from national wealth to parental investment rather than vice versa. Evidently, parents in developed countries invest more in their children's education because doing so is rewarded by offspring success in the job market. In today's competitive skills-based economy, a college degree serves as an entrance ticket to jobs that yield a living wage. Such advantages from education occur only in countries that are already affluent and highly productive.

Workers are more productive if they enjoy good health and lose fewer years of their lives either to chronic illnesses or to premature death. This phenomenon has a simple explanation. Those who are chronically ill cannot put out as much physical or mental effort. The same is clearly true of undernourishment, and these variables are closely linked because poorly nourished individuals are more vulnerable to illnesses.

Living in a country where residents have little confidence in their own future survival has a relevant psychological impact as well. Knowing that the future is uncertain, residents of subsistence societies live in the present: they try to enjoy their present lives as best they can. They discount the future and make little effort toward realizing goals in the distant future, such as saving money to purchase a better home. They have less interest in working hard to accumulate money. They are disinclined to loan money to others because its spending value today outweighs uncertain repayment in the distant future.

Children everywhere share a similar mind-set, which means that they never want to go to bed at night because the immediate joy of playing today is so much more valuable than the distant benefit of being rested tomorrow. Children place so little value in the future that they are unwilling to lend money to each other and require very high interest rates equivalent to about 150 percent per *month* to make a loan.

Between poor health, bad nutrition, a relative scarcity of males, and large extended families, productivity in a poor country tends to remain low. It is stuck in the Malthusian trap of a growthless economy, which, of course, was the standard human condition until the Industrial Revolution about a century-and-a-half ago. Each of the four types of Darwinian competition combine to permit switching over from the near-zero growth that existed prior to the Industrial Revolution to the steady growth that followed it. Of course, that transition could only occur in societies where other conditions were met including cheap energy sources and raw materials, transportation infrastructure, factories, mechanization, and cheap capital. Yet having all these advantages counted for nothing if workers were not motivated enough to remain on task throughout the workday and realize the full profit-making potential of the machines. Indeed, work motivation likely has a greater impact by itself than all of the usual historical preconditions for an Industrial Revolution, judging from the fact that the four Darwinian spurs to work effort account for 60 percent of country differences in productivity.

If good nutrition, good health, masculine reproductive competition, and declining competition within households boost productivity sufficiently, this evidently gives enough thrust to the economy that it escapes the Malthusian trap and rises into the clear air of sustained growth. This reasoning helps to explain why some nations grow so much more rapidly than others. It also helps to account for another curious feature of modern economic growth, namely its progressive nature whereby wealth rises consistently from one generation to the next.

CAN ECONOMIC GROWTH CONTINUE?

Every economy has a cycle of growth followed by recession lasting about five to ten years. Even so, as countries develop, the overall trend for the size of an economy, and for the quality of life, is unrelentingly upward. As a result, affluent countries are more than ten times wealthier than zero-growth ones in terms of the purchasing power of wages. Why does growth keep on going like the Energizer bunny? The impact of each of the four aspects of Darwinian competition that drive productivity gets stronger over time as nutrition and health improve, household size declines, and the population tilts to more males. Can this continue indefinitely? Or will productivity decline, contributing to other drags on development such as pollution and climate change?

All four of the competition-related drivers have gotten stronger since the Industrial Revolution. Nutrition per person increased predictably from

decade to decade. Sex ratios rose steadily with development due to reduced male mortality in infancy and childhood and declining rates of
violent death among adult men.[55] Average household size shrank steadily, transitioning from the extended three-generation families of agrarian
societies, through smaller urban nuclear families, to the predominance of
homes with just one adult in modern cities like Stockholm.[56] Development improves health and longevity thanks to better sanitation, clean
drinking water, public health inoculation programs, improved medical
care, reduced smoking rates, and so forth. So life expectancy at birth is
about twice what it was at the beginning of the Industrial Revolution.[57]
These steady changes drive the continuous rise in productivity and wages
over time. They also suggest that future gains in productivity—and hence
in economic growth—are limited.

Household size cannot fall below one. It also seems unlikely that most
people of the future will choose to live alone, even if solitude proves
illusory in the fully networked world toward which we are lurching. In
that world, most people will be connected to electronic networks in some
fashion for most of the time whether that means updating their Facebook
page, streaming news and entertainment, using GPS navigation, having
home security systems activated, or having their vital signs scanned into a
medical database that issues an alert if medical assistance is needed.

The proportion of males in the population cannot rise very much more
either, unless female mortality were to increase for reasons unknown.
Additional male survival is limited once infant and child mortality decline to very low levels and homicide rates close in on zero relative to the
much more violent world prior to the Industrial Revolution. Further economic benefits from nutrition are also doubtful, considering the rise in
obesity and related illnesses such as diabetes, heart disease, and kidney
failure.

Further improvements in overall health are also limited. Indeed, these
could actually reverse if obesity-related illnesses, like diabetes and kidney disease, were to continue rising into the future. So productivity gains
may slow, or even reverse, in the future.[58] Some evidence suggests that
this is already happening with productivity growth slowing in leading
economies, such as the United States and Japan (admittedly a unique case
due to its rapidly aging population). Some observers question the data,
however, pointing out that in the increasingly informal gig economy,
productivity is hard to measure. Assuming it is real, some of the productivity decline could be due to the Great Recession, where productivity
was dented by low demand for goods and services rather than low work
motivation. As the economy advances, there will be more orders to fill so
that employees work harder and achieve higher productivity. Technologi-

cal development is also likely to boost productivity, most obviously by replacing workers and killing jobs.[59]

Digital technology has been important for only a few decades and is irrelevant to theories of the Industrial Revolution in history. Nevertheless, it emerged as a strong predictor of contemporary productivity, and countries where more residents have Internet access are more productive. This suggests that greater penetration of technology around the globe may increase economic productivity and growth. New technologies, such as virtual reality and the Internet of Things, may help workers to be more productive.

Darwinian competition—as assessed in terms of good health, caloric intake, small household size, and the proportion of men in the population—accounts for most of the country differences in productivity.[60] When these conditions are favorable, workers are more motivated. That a motivated workforce accomplishes more than a lazy one will surprise no one. Yet many scholars would rather focus on other influences, from government to religion, from infrastructure and science to education, and from climate to inflation and taxation. Each of these plausible influences are frequently cited in explanations of why the Industrial Revolution happened (first) in England as well as explanations for why some countries today grow much more rapidly than others. That disparity in growth generates the huge differences in wealth around the globe that garners increasing political attention. Note that annual growth rates are often higher in underdeveloped countries so that China has a much higher growth rate than the United States. In the end, it is not what happens from one year to the next that matters so much as the cumulative impact of sustained development. In terms of national production per person, China is still about four times poorer than the United States.

China's residents suffer from a lack of political freedom. Could that fact hold back their economy? This is a reasonable hypothesis given that countries having the highest quality of life generally benefit from a good measure of political freedom. Given China's rapid growth, this seems unlikely, however.

DEMOCRACY AND ECONOMIC GROWTH

All of the wealthiest countries are democratic, and the poorest provide few guarantees of personal security or freedom of expression. One thinks of North Korea where the sky used to be dark at night everywhere but in the capital city because residents couldn't afford to use electricity. Of course, many were also starving. Anyone who spoke out ended up in a work camp where conditions were even worse.

Its neighbor and primary ally, China, is a great deal more prosperous, but residents have no freedom of expression on political issues there either. All records of the Tiananmen Square massacre—in which peaceful student protesters were attacked by the army resulting in scores of deaths—have been wiped from the Chinese Internet.[61] The government uses a firewall to block references to the atrocity in web pages. Dissenters are arrested without due process and disappear into Stalinist gulags where hard labor is combined with prison guard brutality and a lack of human rights.

Ironically, China was the fastest-growing country in the world over the past three decades averaging about 10 percent per year (although official statistics are skewed by political targets and therefore unreliable). China's brutal One Child policy was recently relaxed but created much misery, particularly in the countryside where extra children were more easily hidden. Such undocumented children could not go to school or get married and were destined for a lifetime of slavery on the farm. With the movement of some three hundred million peasants into towns and cities—the greatest movement of human population ever recorded—fertility has declined substantially, and this likely had a much greater impact on household size than any government policy.

The One Child policy did increase the number of males in the population, however.[62] There is a strong preference by Chinese parents for sons because they are the gender tasked with supporting aged parents and government social security for the elderly is poorly developed. Sex-selective abortions produced a substantial excess of males in the population. With generally good health and nutrition, small extended families, and an excess of males in the population, Chinese workers are highly motivated and their exceptional productivity speaks for itself.

China looks like an exception to the rule that political repression destroys economic growth, but it is worth remembering that its per capita GDP remains a small fraction of that in developed countries. It is a very large, but still rather poor, country. Whether it can develop much more without political reform is an open question, and the centrally planned economy is running into severe problems, not least of which is a very high level of public debt.

Much of that debt was incurred through overbuilding of infrastructure as a way of artificially spiking growth to accommodate the arrival of tens of millions of peasants in Chinese cities each year. These new arrivals leave the subsistence life of peasants and enter the monetary economy, helping to explain the phenomenal level of economic growth.

Many cities have "ghost" neighborhoods with new apartment buildings that are mostly empty and new streets that carry no traffic, a phenomenon pinpointed by researchers who simultaneously mapped the numbers of Baidu users and the number of housing units.[63] The Chinese overinvestment in infrastructure and housing highlights the impact of centralized governments on economic development—good as well as bad. If governments invest in the economy, this gets recorded as growth.

Some social scientists believe that the quality of government is as important as its size and willingness to spend money. The argument is that most of the developed countries today are democratic with the European social democracy as the typical form of government. Is democracy really important as a determinant of economic productivity and growth?

I assessed the strength of a democracy using a comprehensive index developed by *The Economist*. This looks at a whole range of subtopics from having clean elections to being comfortable with hearing political or religious views that differ from one's own. I found that countries with high scores on the democracy index had more productive workers. Yet, from a statistical perspective, knowing about how democratic a country was told me nothing about its productivity that I did not already know from health, household size, nutrition, and the sex ratio. Well-developed democracies have higher productivity because they have better health and nutrition, smaller households, and higher sex ratios that make for motivated workers.

If strong democracies have more productive workers, one would also expect that bigger government would go along with higher productivity. After all, bigger governments can do more to develop infrastructure like roads and railroads, as well as boosting health, education, and research. This view is controversial because most political conservatives would say the opposite, that a lot of government spending takes steam out of the private economy. Their thesis is that government spending is inherently wasteful, and that the same money retained by private owners is more stimulating to the economy.

Yet they are wrong. I measured the size of the government in terms of tax receipts as a proportion of GDP. Countries with high rates of taxation, both personal and corporate, were significantly more productive. Once again, the size of the government in terms of the tax base told me nothing about productivity that I did not already know from health, sex ratios, nutrition, and household size. Countries with high income taxes are likely to have a well-developed welfare state and good public medical services that improve health and survival (particularly for males) and may even facilitate the breakup of nuclear families by reducing the economic importance of fathers, thereby bringing down average household size in modern cities like Stockholm without disadvantaging children.[64]

Most economists believe that an educated workforce would be more productive, and there was a significant positive correlation between average years of education and a country's productivity. Yet I found that such human capital differences cannot explain productivity differences once the effects of health, household size, nutrition, and the sex ratio are taken into account.

Scientific education seems particularly important in modern societies. Scientific or technological breakthroughs are credited with driving economic growth—at least since the Industrial Revolution. Are such advances important for productivity of workers? Surprisingly, the answer was no. The level of innovation in a country—as tapped by patent application rates—did not account for country differences in productivity. This may be because transformative technologies—such as the cell phone—diffuse very rapidly around the globe, producing benefits for those far away from the site of the innovation. Indeed, I found that *use* of modern technology, specifically the Internet, does boost productivity.

No doubt, the same will be true of innovative technologies of the future, including the use of refined robotics coupled with artificial intelligence. The rationale is fairly clear: digital technology and artificial intelligence replace service workers in much the same way that machines got rid of manual laborers. This is an aggressive, ongoing process that is expected to separate at least a third of contemporary workers from their jobs, although such projections may underestimate the number of new human jobs created by innovation given that the technology must be installed, managed, maintained, and updated. None of this would be possible without an infrastructure of cables, satellites, and wireless signals that are as important today as the ships, bridges, and railroads of earlier times.

INFRASTRUCTURE

Surprisingly, the rate of patent applications had no impact on national productivity (once nutrition, health, sex ratios, and household size were accounted for). Actual use of modern technologies (rather than inventing them) may be what affects productivity. Of course, this presupposes an up-to-date information technology infrastructure. It also raises the issue of what role public infrastructure plays in economic development. Investment in physical infrastructure, such as roads and water supply, likely contributes to productivity, but it is hard to analyze due to the absence of a single index (because there are so many different types of infrastructure) and because detailed data are available only for large countries. [65]

The role of infrastructure is illustrated by the consequences of destroying factories and other vital infrastructure during World War II. After the war, the United States benefited greatly from the elimination of industrial competitors around the globe—most of whose factories were bombed—whereas American facilities remained unscathed. Many new factories had to be built in the United States to satisfy global demand. If you must wage war, then destroying the infrastructure of key global competitors, such as Germany and Japan, may bring a huge economic windfall.

For their part, the Europeans cleared out the rubble and rebuilt spanking new factories that enabled much greater efficiency than the creaky infrastructure that was destroyed. This was a period of unprecedented increases in productivity on both sides of the Atlantic, and it did a great deal to establish the wealth of leading countries in the world today (Japan and Germany included).

A country like Russia has much more trouble building a network of roads than one covering less territory, such as Denmark, and it would be surprising if these differences did not affect the velocity of development. Countries clearly differ in the amount they need to spend on infrastructure. All else being equal, countries spending more money on their infrastructure would likely be more productive.

The level of public investment in infrastructure is not tallied directly by most countries. It is reflected in the size of government (measured as taxation level), however. Bigger governments have more money to spend on roads and other public infrastructure. Countries with relatively big government are more productive because Darwinian competition motivates the workforce.

The limited information available (for thirty-six nations) suggests that countries having a greater proportion of their roads paved are more productive, according to my unpublished analysis. That is hardly surprising: workers waste less time commuting and make deliveries more efficiently if there is a decent road system than if they have to use dirt roads that are constantly getting washed out by rivers.

So physical infrastructure does affect productivity and development. Some scholars see the natural physical attributes of a country as the key to its economic destiny. Is the soil fertile enough to support a larger population than those who cultivate the land? Early cities developed in fertile river basins of Iraq, India, China, and Egypt, and agricultural innovations in medieval Europe prompted a repopulation of cities that had declined following the collapse of the Roman Empire. [66]

Agriculture is a tiny fraction of the GDP of developed countries, and agriculture is also a global business thanks to effective refrigeration and other forms of food preservation and storage. With highly efficient modern bulk transportation, it no longer matters very much where food is produced from an economic perspective. Of course, environmentalists argue that local food generates less greenhouse gases, but modern factory farming and transportation are both so efficient compared to the local subsistence farming model that this is evidently wrong. More productive nations are better fed, but where they get their food from is unlikely to affect how productive they are. What about climate?

CLIMATE

For centuries, scholars linked the climate to work motivation. The temperate weather of northern and western Europe supposedly helped people to work harder whereas the hot and sticky conditions of tropical Africa inhibited effort to the bare minimum needed for survival. Yet these ideas did not pan out in my study. I found that temperate countries were not significantly more productive. Moreover, in a climate-controlled world, exterior conditions are increasingly irrelevant.

If temperate countries have no advantage, what about the discouraging impact of tropical heat and humidity? I found that tropical countries were indeed significantly less productive. Yet the disadvantage was due to disease rather than climate as such. When health status of a country (in terms of disabling chronic illnesses and early death) was statistically controlled, tropical climate had no effect on productivity either.[67] Work effort in tropical countries is undermined by poor health rather than by climate as such. Productivity is lower in some tropical countries because they have a heavier burden of disease—not because the climate is uncomfortable. Indeed, some of the most developed—and therefore productive—countries, such as Hong Kong and Singapore, actually have tropical climates. They have state-of-the-art medical systems that enable residents to lead long, healthy, and productive lives, however.

Geography need not be destiny. That is why organizations such as the Gates Foundation are committed to eliminating some of the most disabling chronic illnesses, like malaria, from sub-Saharan Africa and other tropical regions of the world. They believe that if health can be improved, economic growth will follow, permitting these disadvantaged nations to solve their own problems in the future. Of course, this is exactly the opposite of what Thomas Malthus believed.

According to Malthus, the only hope for an improved quality of life was population reduction whether by epidemics or by warfare. According to Malthus, improving health and increased population inevitably exhausted the food supply and reduced the quality of life. Most people find this thesis unbearably depressing today, but it was not wrong in the past.

Modern societies benefit from green agriculture that raises the food supply in ways that Malthus could not have anticipated. Scientific agriculture permits the food supply and population to increase in lockstep in a manner that Malthus could not have imagined.

Climate, as such, is not a major determinant of contemporary economic productivity. Thanks to climate control and improved medicine, it has minimal effects on worker motivation. Workers become more highly motivated when the benefits of unusual diligence exceed the costs. This happens when Darwinian competition over food, household resources, and survival is reduced but men must earn more to enhance their appeal to potential partners.

THE MOTIVATED WORKER

Improving health is a major ingredient in the rising productivity of developing countries. Healthy people work harder because they are able to and because they want to. They are motivated to work harder because they anticipate a long life and a better future. They want to save money to make that future secure. Conversely, residents of underdeveloped countries, whose life expectancy is low, focus on enjoying whatever pleasures are available now and do not put much stock in the uncertain future.

Better health means that more and more males survive in developed countries. This drives up competition for the affections of women and provides men with an additional motive to work hard and achieve a financial status that makes them more desirable as mates. Countries with a higher proportion of males are indeed more productive. As the sex ratio rises, women, as well as men, become more productive. Men work harder due to increased competition over women, but why are women working harder in modern economies?

The answer to this question may be found in the related question of why female labor participation increased so sharply in the second half of the twentieth century. In other words, if you know why the world of paid work become so much more attractive to women, then you can better understand why women want to work harder and be more productive.

The impetus for many women entering careers in the 1960s was a temporarily poor marriage market wherein single women at the peak ages of marriage outnumbered single men by as much as five to four in the United States and other post–World War II countries.

The number of married women at work took off during the 1960s, whereas most wives of the 1950s had stayed out of the workforce and focused on raising their children. What changed? With the expansion of the service economy, many more jobs were available for women whereas there were fewer jobs for men and their earning power suffered with adverse consequences for household finances. In particular many of the well-paid semi-skilled jobs for men disappeared with advancing automation.[68] So the one-wage family of the 1950s was replaced by two incomes, often in low-paid service jobs. Married female workers were partly motivated by economic necessity, but rising standards of living were also an issue.

Once married women began working in large numbers, the trend kept going for two good reasons. First, in an age of television marketing, ordinary people were suddenly made aware that their standard of living was mediocre compared to that depicted on TV shows and they aspired to live more luxuriously. Second, and more important, the cost of living increased due to inflation in many of the expenses related to raising children.[69] So the mother's income was increasingly important.

Women became ever more ambitious and economically successful. One measure of female career motivation is soaring enrollments in college education that rose to exceed male enrollments by the end of the twentieth century. So there are plenty of reasons why women became more ambitious, more hardworking, and more productive. This boosted their productivity.

The connection of Darwinian competition to productivity is complex. People become more productive when Darwinian competition is relaxed as to health and nutrition. Declining competition among extended family members also liberates individual ambition from the smothering embrace of the old three-generation extended family that provided a haven for lazy freeloaders at the same time that it discouraged taking responsibility for the future or working toward a better life. Such a family structure was tailor-made for the Malthusian world prior to the Industrial Revolution but was an early victim of that transition. In the modern world, social mobility is not only possible but expected, and social comparison (that is facilitated by electronic media) motivates people to earn more. In this world, a man's earning capacity is key to his romantic appeal, and the fact that more males survive exacerbates the competition to appeal to women. This induces them to work harder. Whereas several types of Darwinian competition are reduced in highly productive countries, male mating competition increases.

COMPETITION IS THE KEY TO MODERN PRODUCTIVITY AND GROWTH

Productivity rises with improving technology. That principle is particularly salient today. Fewer and fewer employees get the same amount of work done thanks to information technology. Yet technology and infrastructure pale in importance compared to an employee's interest in working hard and maximizing output. Take the low productivity of Indian machine looms early in the twentieth century, for instance. Of course, it was not purely a matter of work motivation. Workers were also unaccustomed to running machines causing frequent breakages and stoppages. Indian managers were poorly versed in efficient work practices, such as using teams of specialists (doffers) to remove empty yarn spools from the spinning machines thereby minimizing downtime.

Compounding a general lack of familiarity with machines, workers in less-developed countries are less acutely conscious of time than is true for societies where most people carry watches. If they have a poorly developed precise sense of the passage of time, they are less concerned about how much product they generate each hour. So productivity is complex. Yet most of the international differences in worker productivity (measured as GDP per hour) are explainable in terms of four measures of Darwinian competition.

The simplest of these measures involves the health status of the population. People can't work if they are weakened by chronic diseases such as those that hobble productivity in tropical countries lacking a well-developed health-care system. Malnutrition saps work motivation for similar reasons. Biological vigor, or the lack thereof, prevents residents of impoverished countries from working hard, and this holds back their economies, just as inability to obtain food and resist disease cuts the reproductive success of animal populations.

Members of subsistence societies discount the future and pursue current pleasures. They try to make the most of their current lives by laboring as little as possible and are not interested in working toward a higher standard of living in the future. The future is highly uncertain given low life expectancy, disease epidemics, and high levels of interpersonal violence—which is why it is "discounted." Given this uncertainty, there is a prevalent unwillingness to lend money that drives up interest rates and cramps business activity. This phenomenon recalls the principle that in animal populations experiencing heavy predator pressure, prey species play it safe, seldom moving far from refuge.

Countries having large extended families are less productive also, an effect that psychologists refer to as social loafing. One version of social loafing was recognized by Russian economist Alexander Chayanov early in the Soviet era. He concluded that agricultural workers in Soviet collective farms were motivated solely by basic survival needs. Once those needs were met, they loafed. Doing any more was futile because the proceeds of greater industriousness were distributed among other members of the collective. Of course, most social animals seek safety in numbers where they pool security measures, territorial defense, and knowledge about the location of resources.

Finally, in societies where proportionally more males survive due to better health care at all ages, there is greater male competition for women that gets expressed in economic competition given that women select higher earners as husbands. A similar phenomenon is found in other species of vertebrates. Where there is a relative scarcity of females, males invest more in raising offspring. In birds having sex role reversal, where males do more care of offspring, for example, there is generally a substantial excess of males in the population.

Men compete with each other by working hard so as to earn more money than their rivals. Financially successful men are more attractive to women, who mostly prefer working men to the unemployed and white-collar workers over blue-collar ones. The sexiness of money was established in many studies by social psychologists that include the Internet age in which far more women join the workforce and achieve the highest levels of occupational success.

Women are not only more ambitious, earning more bachelor's degrees, and more active in the labor force today, they are also as productive as men despite the fact that women do not make themselves more attractive to men by earning more.[70] Countries with more women employed are not less productive (or more productive). Recruitment of women to the paid labor force from the cross-gender cooperation of subsistence societies is one of the most consequential economic changes of the twentieth century that is intimately tied to economic development.

WOMEN SHIFT FROM UNPAID TO PAID WORK

Women have always worked harder than men. This is true whether they are judged by the time devoted to productive work or the amount of productive activity accomplished in a day. This conclusion emerges from the time budgets of hunter-gatherers where women generally work for longer than men each day. In all societies, women perform most of the domestic work and do most of the child care. In subsistence societies, they also gathered food, tended crops, and looked after domestic animals.

Important as these activities were, they were conducted outside the monetary economy and had no impact on gross domestic product.

Much of that "free" labor got shifted to monetary activity in the service economy that really took off in the middle of the twentieth century. With plentiful job opportunities, most married women work for pay in developed countries in addition to doing the lion's share of unpaid work in the home. What motivates them?

Highly productive societies generally have a high, or rising, cost of living, including high costs of raising children. If household costs go up sufficiently, two incomes are required for the family to keep its head above water. The problem involves not just inflationary forces in general but the cost of raising children in particular. We see this reflected in home prices where densely populated modern cities like Singapore raise very few children for economic reasons. Of course, home prices are an essential part of the expense of raising children given that parents starting a family generally move to more spacious accommodation.

Wanting a higher quality of living also induces women to work harder and pursue more ambitious careers. Of course, this reasoning applies to men also. When television first arrived in the Western states, men as well as women suddenly felt dissatisfied with the drab quality of their daily lives.[71] Women may have felt this materialistic envy more keenly because they are the ones making most of the purchase decisions in a household and are therefore more acutely aware of the material quality of their homes. Moreover, women used to have exclusive control over household arrangements so that a shabby environment reflected negatively on women in the household rather than men. Perhaps for that reason, women seem more likely to notice the torn window blind or the worn rug that bespeak poverty or neglect.

Whereas men may be more interested in products that signal high social status, such as sports cars, boats, and Rolex watches, women are more interested in owning expensive items that they may wear, such as expensive jewelry, fur coats, and luxury brand bags, like Coach or Michael Kors.[72] Luxury brands of wearable items are of interest not just to affluent women but to those having moderate incomes, possibly reflecting the greater importance of physical appeal for women than men.

The reasons that women are more productive in developed countries are quite transparent. Like men, they are becoming healthier and more future-oriented. They live in smaller households where the fruits of their labors go just to them (in the case of single women without dependents) or just to their immediate nuclear family (in the case of married women)—powerful motives in modern economies where raising children is extremely expensive. As the gender more interested in (or responsible for) shopping, women are more attuned to marketing and more interested in luxury goods that are either worn or displayed in the home. Each of these motives motivates participation in the monetary economy even as their domestic work declines with smaller families and labor-saving gadgets. Improving living standards following the Industrial Revolution arguably played a key role in increased productivity for women as well as men. In this milieu, basic Darwinian competition over food and survival was reduced. These pressures had prevented growth in Malthusian societies. One type of competition increased, however. Male competition over women increased as more men survived to adulthood. This phenomenon actually spurred masculine ambition and work effort.

DARWINIAN COMPETITION AND THE INDUSTRIAL REVOLUTION

Historians see the Industrial Revolution as a product of its initial conditions. In the cloth industry in England, these included cheap raw materials, abundant local coal as the fuel for steam engines, machine looms and an accretion of technological innovations, a well-developed transportation infrastructure, extensive foreign trading networks, and an abundance of cheap capital. Each of these factors may have been necessary for the Industrial Revolution to happen. Their combined effects were not sufficient for growth to occur, however. More than anything else, the sustained increase in productivity that defines the Industrial Revolution requires motivated workers. Motivated workers are time-conscious and stay on-task throughout the workday, doing their utmost to maximize output. Otherwise—as illustrated by the Indian cloth industry early in the twentieth century—the benefits of technology and infrastructure go for naught.

Workers suddenly became more motivated and more productive. Why that happened is something of a mystery to economists whose view of human beings as driven primarily by financial gain, and the goods or services that money may buy, is a Frankenstein-like caricature. Of course we are a lot more complex than that. We are products of evolution more than of markets and our needs precede—and exceed—mere money. This means that any pronounced change in worker motivation is most likely connected to different facets of Darwinian competition, such as that over survival, status, and reproductive success.

Some economists would argue that if British textile workers were so much more productive than their Indian counterparts, then they were simply better managed. Yet that begs the question: why were Indian managers so much less ambitious, or less effective, than their British counterparts? In reality, Indian managers were well aware of the problem and wanted to correct it, but they were confounded at every turn by the inertia of employees who were undisciplined, had a poorly developed sense of time, made little distinction between their private lives and the workday, and spent most of their work hours just killing time. So members of their extended family showed up at the mill to bring them food. They took informal breaks to wash, shave, or drink alcohol. Workers were also rather careless in their treatment of complex machinery (to which they were unaccustomed outside the plant), resulting in numerous breakages. The low productivity of factories was not just a problem for India but is characteristic of all underdeveloped countries where the cheapness of labor does not guarantee profitability.

If you want to understand work motivation, you must recognize that some aspects of motivation are built into us by our evolutionary history. An animal that is starving conserves energy by moving as little as possible. So workers are more productive in countries where more food is consumed each day.

In a similar vein, animals that are sick withdraw from social interaction and rest up to conserve energy. In human terms, poor health translates into absenteeism and low productivity. Being sick also intensifies the focus upon the present and the desire to reduce pain and increase pleasure. People put out less work effort when they are sick because the body needs all of its energy stores to fight illness.

Even if they are not currently sick, people in subsistence societies are much more focused on the present than the future. It is as though the future has little reality, so that they do little planning or preparation for the unreal world ahead and live in the present. (Of course, farmers are forced to be more future-oriented than hunter-gatherers because a crop that is planted today may not be harvested for many months.)

In developed countries, the future looms larger. Many workers save for retirement decades ahead. Current health and future orientation are key components of work motivation, but there is at least one more key evolutionary component—the difference between busy grazers and lazy predators.

Herbivores spend most of the waking day either consuming plant food or digesting it (as in cows chewing the cud). They are active throughout most of the waking day so as to obtain sufficient nourishment from a diet that is low in energy, high on fiber, and difficult to eat and digest. On the other hand, carnivores are notoriously lazy, so that a snake might consume a large meal and sleep for months. Even mammalian predators, such as lions, that have higher metabolism and need to eat much oftener, are notoriously lazy and typically sleep for as much as twenty hours per day. Similarly, human hunter-gatherers spend much of their waking day resting or socializing. Our ancestral lifestyle approximated that of carnivores rather than herbivores (although women and men collected vegetable as well as animal foods).

Carnivores consume a much more energy-dense diet so that they need to be active for less of the day and they conserve energy during their "down" time by resting or sleeping. So the human transition to agriculture was difficult because our ancestors were unaccustomed to long hours of arduous, repetitive toil. Their bodies were not equal to the strain, and they paid a price in terms of repetitive stress. Work-related deformity of bones and joints first entered the fossil record following the Agricultural Revolution.

Long hours of toil were difficult and unnatural for our ancestors. We are, after all, designed more as easygoing predators than as hardworking herbivores. If people were forced to work long days for whatever reasons, one might expect them to take frequent rests, as any predator would. This is consistent with anthropological evidence on hunter-gatherers. Agriculturalists worked considerably harder but resisted hard labor if they could. Loafing is a well-known problem in developing countries where factory owners find it more difficult to turn a profit than in developed countries where wages are considerably higher.[73] Then, preceding the Industrial Revolution, there was a spurt of industriousness that served as a key driver of modern productivity and wealth.

So why did English workers become markedly more productive after 1860 even compared to the Industrious Revolution that had gone before? A key consideration was improved health thanks to antiseptic use and other advantages of more scientific medicine that began around 1850 with the first use of carbolic soap that was used as a surgical antiseptic and reduced deaths from childbirth. Average life expectancy increased from about forty years in 1860 to about seventy-eight years in 2000.[74] With improved health, workers were better able to work. Longer life expectancy made employees more future-oriented because there was more of a future for them to look forward to. So they were interested in saving to ensure that their future needs were taken care of. Living in smaller nuclear families rather than large extended families meant that workers were more ambitious because increased earnings improved their quality of life and social standing and were not simply absorbed by the extended family, as in earlier times.

Once the Industrial Revolution took off, productivity increases drove up wages and the standard of living rose. This introduced the materialistic competition that drives consumption in modern economies where luxury brands symbolize success and happiness.

Yet none of this would count for much if there was a truly oppositional relationship between workers and capitalists. Opposition is suggested by terms such as "robber barons" that referred to early industrialists in

America such as Andrew Carnegie and Cornelius Vanderbilt. Yet these industrial titans presided over a substantial increase in wages and living standards. According to science writer Matt Ridley, they were really "enricher barons" because they brought down prices of basic materials, increased productivity, and boosted wages.[75] So the wealth generated by increased efficiency of industrial production boosted the living standards of workers to a degree never before seen. This meant that less of their income was required for food, shelter, and other subsistence needs. With cheaper food in terms of hours of work needed to put food on the table, nutrition improved, making employees more energetic and productive. More could be spent on nonessentials such as fashionable dress, books, fine furniture, and leisure-related goods and activities.

Yes, the conditions in a Carnegie steel mill were appalling and dangerous by modern standards. Why did workers risk their lives each day? Clearly, they did so because their wages permitted a substantially better lifestyle than they had experienced scratching out a living on subsistence farms. There was also an unwritten understanding that if a business did well, its workers could be better paid. So the interests of the owners and the workers were more aligned than otherwise. This fact is often lost in descriptions of the famous Homestead strike of 1892 in which Carnegie Steel put down a strike using armed private police (the Pinkertons). History records this as a victory for brutal repression of workers and the successful breaking of a strike. Yet it is helpful to appreciate how this strike arose and how it was perceived by contemporaries.

The strike began upon news that Carnegie Steel profits had soared but that wages would not be increased to reflect this. In other words, workers were upset because they were not receiving a fair share of the profits they had generated by the sweat of their brows. So there was an expectation that profits from the enterprise should be directed to workers to some degree. The conclusion of the story is also revealing. Carnegie may have won the battle but he lost the war. His reputation never recovered from the stain of the Homestead strike and its brutal suppression, and he was forced out of the steel industry.

Why did workers become more productive? It is clear that there was a combination of carrots and sticks. The carrots were good wages and the prospect of earning more over time and building a better future. The sticks were owners and managers who maintained iron discipline, insisting on long hours, with employees remaining on task, and pushing up production. Such enforcement of discipline was necessary early in the Industrial Revolution as former agricultural laborers who had danced to the tune of soil and seasons were molded into creatures of the time and productivity regime of industrial barons. In the long run, though, such time discipline became ingrained in the entire society and incentives grew more important.[76] People worked harder mainly to boost their wages and win a brighter future.

Social mobility had actually preceded the steam-engine-based Industrial Revolution. In the distant past, wages were so low that there could be little mobility: wealth was, and was perceived to be, hereditary. This changed before the era of the factory. Cloth manufacturers "put out" their orders, or subcontracted them, to cottage weavers who worked at hand looms. Using tallow candles that shed a bright and steady light, cotters worked night and day to fill their orders and saw a modest increase in their standard of living. Now, with increased productivity and rising wages, a person could save to buy a better home, better furniture, or more elegant clothing. With the prospect for actual social mobility, every person was responsible for their own station in life and keeping up with the Joneses was born. This period preceding factory production of cloth is aptly named the "Industrious Revolution." It was accompanied by slow but steady economic growth and rising discretionary spending.

CONSUMERISM

Such is the dynamic of rising living standards that the beneficiaries are never really content with where they are. There is always someone else who has a better home, a better car, or better furniture; who sends their children to better schools; or who spends more on cosmetic surgery. Aspiring to lead a more luxurious life is a treadmill: it keeps consumers in debt and constantly working harder to get where everyone else seems to be. (Resentment for those who would rise above us, and the desire to be as good as anyone else, dates back to the era of hunter-gatherers, reflecting the competitive side of human societies.) This is a positive climate for businesses.

The desire for more goods and services is enabled by modern communication technologies. This played out following the introduction of television to towns in the U.S. West, as noted earlier. The increased theft that followed was motivated by dissatisfaction of viewers with their standard of living compared to that depicted on the small screen. In marketing terms, demand for many consumer goods was created overnight.

The same phenomenon occurs on the Internet, where users are exposed to luxurious living by celebrities around the globe. Of course, Internet shopping is the ultimate enabler for luxury goods because these are rarely more than a few clicks away. Consumers want to earn more money so as to purchase more luxury goods ranging from clothing to travel. The carrots of consumption may be more potent than the sticks of management in boosting economic growth. If so, then the Internet, which encourages spending like no medium before it, is good for productivity not just because it drives efficiency (by replacing service workers) but because it spurs consumption as well. Modern societies are thus something of a rat race where people work hard at consuming during their off time and work even harder during their hours on the job so that they can earn enough to pay their bills.

We are accustomed to thinking of the Industrial Revolution as occurring in factories, but the critical changes were likely psychological rather than technological. People "suddenly" became willing to work harder and produce more income than what was required to meet their survival needs. Scholars are divided about exactly when worker productivity jumped. There was a gradual change between 1760 and 1860, but the increase in productivity after 1860 was much more pronounced, coinciding with improvements in health. Increased productivity accounts for rising wages. Higher earnings facilitated the rise of consumer societies where workers are motivated more by luxuries than by necessities.

Lurking inside the transition to a more productive workforce was the realization that the success of the enterprise was critical to the interests of the workers, whether in terms of job security, improved wages, or both. The history of the industrialization period is replete with accounts of brutal working conditions and bitter conflict between management and workers. Yet the lives of factory workers must have been better than those of agricultural workers, or workers would have remained on farms (at least until definitively replaced by mechanization early in the twentieth century). Tartars such as Andrew Carnegie notwithstanding, industrialists grasped the benefits of paying their workers well to ensure a better, more motivated workforce. Perhaps the clearest example of this principle was Henry Ford, who deliberately paid employees enough so that they could buy his cars even though these wages were at a premium to the market. Of course, Ford knew that the automobile was the must-have consumer good of his day and that the "surplus" wages would come right back to him in sales.

Despite the bad press received by robber barons, business owner interests have always been inherently aligned with the interests of workers. A profitable enterprise can pay higher wages and often willingly does so if it produces superior returns in terms of employee skills, loyalty, and productivity. This phenomenon is illustrated by the rise of large corporations, such as Microsoft, Google, or Apple, where explosive growth in profits and market capitalization are tied to enrichment of valued employees. Many of these are turned into millionaires thanks to generous salaries, retirement benefits, stock options, and other perks.

The success of high-paying companies would suggest that treating workers well can pay off for the business, and this is certainly true of industries such as software, electronics, and the Internet where the specialized skills of a relatively small number of people makes the difference between success and failure. Even in the early days of the Industrial Revolution, top workers must have been motivated by an awareness that what they did generated real value for the company and effective managers got the best out of their skilled employees by nurturing and rewarding their talents.

This was certainly true of engineers who built the machines that replaced human (or animal) exertion thereby increasing efficiency of production by orders of magnitude. We know that foreign governments, including France, Germany, and the United States, courted English engineers so as to steal their secrets or even hire them away to found their own industrial economy. By the heyday of Andrew Carnegie, there was a widespread expectation that increasing profits of the enterprise should get fed through to the workers as higher wages. That sort of environment is very favorable to increasing worker productivity given the expectation that however hard they work, they will be amply rewarded. Carnegie mastered all other aspects of efficient steel production, but he did not appreciate the delicate unwritten social contract between owners and workers such that the fruits of their labor should be partially returned to them if they are to keep with the program and continue to do their best work.

English workers endured rather inhumane conditions early in the Industrial Revolution with long hours, noise, air pollution, and constant risks of accidents on the job. Even so, they were better paid than agricultural workers and could save some money. Of course, the entire period since 1860 has been one of steadily rising wages and improved living conditions, a trend so powerful that even economic catastrophes, such as the Great Depression, show up as a minor blip on the historical chart. Rising wages fueled consumerism as an ever-smaller portion of worker pay went to necessities like food and shelter, thereby increasing disposable income and leisure pursuits.

The improving quality of life was reflected in increased longevity. Steady increase in average heights of European military recruits implies that nutrition was improving along with wages. Between 1870 and 1970, heights increased by an average of 1 cm per decade with the greatest increases occurring between 1911 and 1955, a period that included two world wars and the Great Depression but also experienced major improvements in public health and sanitation. Interestingly, the century preceding 1860 (sometimes referred to as the "first Industrial Revolution") was characterized by a marked *decline* in stature. During this period, manual work was getting replaced by water mills and steam power and many agricultural workers relocated to cities. The stature data suggest that this period of transition was hard on worker populations. One plausible explanation involves growing up with increased exposure to epidemic diseases in cramped urban homes with poor sanitation and polluted air. Even if wages were rising, the transition to urban life posed serious health challenges.

Rising wages and the accumulation of wealth by workers meant that ordinary women became more conscious of money issues in marriage. In the feudal age, all peasants were equivalently poor, scratching out a living on land they didn't own. With increased prosperity, distinction based on wealth was much more important. It was as though the working class had acquired the lower-gentry concerns of characters in a Jane Austen (1775–1817) novel. Women focused more on the economic prospects of potential husbands. Men who had bleak job prospects, or who were poorly paid, were at a disadvantage in the marriage market. As the supply of men increased, thanks to improved scientific medicine that saved more males (who are more vulnerable to childhood diseases than females are), competition between them intensified.

Hence the emergence of an intensely class-conscious society that emphasized fine distinctions between the status of a messenger and a footman, or between a footman and a butler. These fine distinctions had become much more important because they meant the difference between marriage and nonmarriage. For those men who married, good financial prospects meant higher mate value of the bride. Money attracted women who were physically attractive, agreeable, socially skilled, and well connected (i.e., Jane Austen heroines). Social mobility and successful marriage motivated men to work harder. Rising class consciousness motivated men and women alike to spend more on luxury goods that expressed their relative status. These nineteenth-century concerns are less important today because we live in a society where hereditary rank is relatively unimportant. Yet how much money people have still matters a great deal, and living well is the ultimate proof of social success. Hence the unbridled consumerism of the modern era. The orgy of spending that economists see as very good for the global economy, environmentalists see as very bad for the planet.

Growth of the world economy over the past two centuries produced a staggering increase in global trade that now accounts for most of the products consumed within a country. This is very different from agricultural societies that mostly consumed their own products. Ever larger, more efficient, and more numerous cargo vessels transport goods ever more cheaply around the globe, facilitating rapid growth in trade. Apart from the sheer volume of goods transported that generate huge islands of plastic waste in the world's oceans, the world economy has integrated supply chains. So an iPhone sold in the United States that is nominally the product of China has components made in the United States itself in addition to Singapore, South Korea, and other countries. This modern way of life is very different from the low-impact lifestyle of foragers who moved camp often, leaving the habitat pretty much as they had found it. Needless to say, these ecological changes posed major adaptational problems for our species. The final chapter looks at some of the ways in which human psychology and behavior adapt to the contemporary world.

7

HOW HUMANS ADAPT TO THE MODERN ENVIRONMENT

How did our ancestors get themselves out of the zero-growth Malthusian funk of the past? That is a question that economists still struggle to answer. Our trajectory changed dramatically with the Industrial Revolution. Many preconditions had to be in place for the Industrial Revolution to occur, such as new energy technologies, mechanized factories, concentrated capital, availability of raw material, and access to markets. Yet the human elements of work motivation and need for achievement are more important than is generally appreciated.

This is a complex phenomenon. Human motivations affect the economy, but the economy also affects human psychology. To begin with, there is an important motivational element underlying economic development. Economic similarities in modern economies also drive many convergences in social behavior, from reduced fertility to rising gender equality.

Such profound adaptability of the human species is commonly assumed to be unique. Yet it is wise to keep an open mind. Is the human transition from Malthusian stasis to economic development any more profound than the transformation of solitary grasshoppers to swarming locusts following rainfall in the desert?

The key transition of the Industrial Revolution was an increase in the value of what a worker produced for each hour of his or her labor—a transformational change because it opened the door to economic growth and modern affluence. [1] Productivity was not just higher before industrialization than after; it has continued its increase ever since. This permits the accumulation of wealth, rising living standards, and many correlated changes from bigger homes, longer lives, smaller families, and atheism, to epidemics of clinical depression and obesity. In addition to the four

Darwinian spurs to work motivation, technological innovation increases productivity and raises living standards.

MODERN TECHNOLOGY AND PRODUCTIVITY

This book throws cold water on the theory that technological innovation is the primary driver of productivity increases following the Industrial Revolution. Yet new technologies *are* important, and my research suggests that this is as true in the Internet Age as it was in the Age of Steam given that country productivity rises with Internet connectivity, even with Darwinian competition statistically controlled. One way of thinking about technological innovation is that while new technologies give a boost to economies, this washes out over time. From that perspective, many contemporary economists question whether the Internet can be a primary driver of economic growth in the future. With continuing Internet adoption in many of the world's economies, Internet access predicts contemporary productivity. We can anticipate many supplementary effects of innovations in artificial intelligence, virtual reality, the Internet of Things, and so on.

With the widespread introduction of mechanization to U.S. manufacturing industries, from the 1920s on, there was a huge increase in output per worker in the relevant industries. One example was Henry Ford's introduction of the assembly line to car manufacture. Yet such advances generally peter out soon after they are implemented, and the productivity advantages get wrung out. After the assembly line, the next really big advance in car assembly was more than half-a-century away with the introduction of robots that replaced auto workers in a rather ominous sign of the future world of work.

Internet-driven increases in productivity are prevalent and continuous. Digital technology delivered huge productivity gains beginning in the 1990s. Between 1995 and 2004, U.S. labor productivity increased at an average annual rate of 3.26 percent.[2] This means that at the end of this ten years, a worker was producing 38 percent more per hour than they had produced at the beginning. Between 2004 and 2014, productivity slowed to an average of 1.44 percent, thanks partly to a worldwide recession that occurred at the end of the period and also to an aging workforce that has fewer people of working age and also fewer workers of prime spending ages that correspond to the raising of children.

Such fits and starts are an inevitable feature of economic growth everywhere. To some degree, this reflects the discontinuous nature of major disruptive technologies. There is also a statistical aspect: if an economy has grown sharply for several years, it is hard to maintain growth at the same level because the economy has grown so much that the amount of increase required to maintain growth at the same nominal value is much larger. In China today, growth of 6.5 percent is considered low compared to growth rates of 15 percent a decade ago. Ironically, China is actually growing more than it did a decade ago because the economy today is more than twice as large as it was ten years ago. Similarly, the decline in productivity growth in years following major new technology deployment is something of a statistical illusion because the size of the economy has increased. Nevertheless, the easiest gains from innovations are realized early on. The low-hanging fruit getting collected most easily.

However one interprets it, the recent decline in productivity growth rates is an objective fact of the United States in the twenty-first century. Pessimists see this as a sign that our economy is on the cusp of serious secular decline. Yet one should not make too much of short-run patterns. This would be like arguing that the Great Depression would bring about a subsequent collapse of the global economy. In reality, the end of the depression augured a global expansion unlike anything ever seen before. Economists believe that deep recessions provide opportunities for new entrepreneurs and for innovation by shaking out the dominant players (an idea developed by Joseph Schumpeter).

Even if current problems are nothing like the Great Depression, we have gone through two major financial crises in the new century at a time when the benefits of initial adoption of digital technologies, such as electronic records, cell phones, and online commerce, are waning. The easy gains from digital technology for businesses (e.g., hiring fewer sales-clerks due to Internet commerce) have already been realized. Nevertheless, many new disruptive businesses are on the way from smart cars to smart homes and from virtual reality gaming to drones as replacements of workers that range from delivery people to soldiers and cowboys. Such rapid transformations are illustrated by apps such as Uber replacing the conventional taxi industry in a few years and apps such as Airbnb making inroads into the hotel industry. Considering the many technological innovations in the pipeline, we should hesitate to claim that productivity cannot continue to grow.

Futuristic applications already in development include interacting with electronic devices by gesture and through voice so that our communications with machines become much more natural. Remote control of domestic devices means that homeowners will easily monitor and change settings in their homes while on vacation. We are entering a new era of much greater connectivity between electronic devices in the Internet of Things. Critics of recent advances in digital technology point out that many increase the efficiency with which we use our leisure time rather than what we do at work. Examples include Facebook and Snapchat, which are good at sharing information between (actual) friends but may reduce workplace productivity by wasting time.

A similar argument could be made about widespread adoption of leisure time reading in history. Yet it would be absurd to claim that an illiterate country can be more productive than a literate one, and the same is obviously true of digital literacy. Improving digital technology holds enormous promise for increasing productivity and affluence in the real world, which is why Google and other large tech companies are pushing ahead with Internet access projects in underdeveloped countries.

The Industrial Revolution is far more dependent on psychological change than most economists would credit. Work motivation increases productivity, but economists historically underestimated psychological influences because their discipline focused elsewhere on land, labor, capital, technology, and raw materials. For economists, labor is defined more by cost and level of skill rather than by work motivation. Even so, huge differences in productivity in countries around the world that are poorly explained otherwise can be explained in terms of varied work motivation. This book presents new evidence that how hard people want to work is related to nutrition and health in addition to other aspects of Darwinian competition, including male sexual rivalry and conflicts among members of extended families over how resources are distributed.

Technological development may increase economic growth and general affluence, but it is rarely an unmixed blessing. Following the Industrial Revolution, cities provided a less than optimal environment for workers. There was an initial decline in nutritional quality (as indexed by reduced stature) and increased vulnerability to infectious diseases due to crowding as well as respiratory diseases due to pollution.[3] Of course, similar problems arose following the Agricultural Revolution as diets became less varied, combined with periodic famines. Agriculture increased population density, providing a reservoir for infectious diseases, as illustrated by high mortality from diarrheal diseases in poor agricultural countries like Guatemala today. Such diseases have largely been vanquished by improved public sanitation and scientific medicine in more developed countries. The nutritional problems from early in the Industrial Revolution have largely been addressed thanks to improved overall living standards.

Ironically, the dietary problems of the Agricultural Revolution are being relived today in the sense that processed modern foods are of lower quality than a hunter-gatherer diet, and many nutritionists worry that there is too much emphasis on a few staples such as corn, rice, and wheat. Overuse of corn-based sweeteners is believed to constitute a health threat in terms of obesity, diabetes, and related illnesses. Similarly, trans fats used in cooking are blamed for clogging arteries, so they were banned in restaurants in New York and other cities.

Far from being entangled in extended family obligations, modern workers mostly live as singles, as couples, or in small nuclear families. In some highly productive cities, like Stockholm or San Diego, the majority of workers are single and live alone. Hence the phenomenon of young go-getters devoting their entire waking lives to prominent companies like Apple, Facebook, Baidu, or Google that are founts of innovation and productivity.

Unfortunately, the very success of such companies means that users spend ever more of their lives in sedentary pursuits, perhaps working from home, shopping online, using dating sites, updating their Facebook profile, talking with and texting their friends, sharing pictures, or watching Netflix. This sets them up for the peculiarly modern health problems of obesity and associated diseases that are now the greatest threat to leading a long and healthy life. Given that humans are predicted to respond adaptively to environmental changes of this sort, what can we make of this seeming adaptive failure?

THE OBESITY EPIDEMIC AS AN ADAPTIVE PHENOMENON

Obesity is an unmistakable problem of modern life. Look at a picture of a hunter-gatherer community such as the Ituri pygmies featured in the Jim Carrey movie *Ace Ventura Pet Detective*. Everyone is completely toned and fit and there is zero obesity. Now look at a picture of the Pima farmers in the mountains of Mexico. Once again everyone is skinny as a rail thanks to constant physical activity. The Mexican Pima drew attention from obesity researchers because their American cousins living in Arizona were identified as the most overweight ethnic group in this country. This highlighted the important role of environmental factors (rather than genetics) in modern obesity.

When people in developed countries see pictures of skinny foragers and farmers, they may assume that the subjects are slim because they are undernourished. Yet anthropologists report that indigenous people do not eat less than members of developed societies. Indeed, adjusted for body weight, they eat far more than we do for the simple reason that they expend far more energy in everyday activities.[4] An Ache forager from Paraguay consumes approximately twice as much food as a modern American (when appropriately adjusted for body weight) and subsistence farmers like the Mexican Pima are also well-nourished. So for humans with the naturalistic activity levels of subsistence societies, it is almost impossible to become obese, and almost no obesity (or overweight) is observed in populations where everyone performs manual work.

For mostly sedentary populations, such as ours, matters are very different. Average levels of caloric intake can result in obesity for vulnerable individuals. Inactive populations are also more vulnerable to feelings of hunger. The underlying physiological reason is that prolonged activity mobilizes fat stores thereby suppressing hunger. Moreover, indigenous peoples do not suffer from the insulin resistance that is a major cause of obesity in developed countries. Sedentary populations eat more frequently, and snack foods are typically high in fats and sugar. When experimental rats are inactive and exposed to a "supermarket" diet, they also eat too much and gain weight.[5]

The modern epidemic of obesity is a predictable outcome of leading a sedentary life. That fact is obscured by proponents of various weight-loss products including fad diets, appetite suppressants, and even surgery to reduce stomach size. None of these approaches is generally desirable, and all have the potential for adverse side effects. What is needed instead is about two hours per day of moderate physical exercise, such as walking a distance of eight miles. This happens to be approximately seven times the activity level recommended by physicians. This is based on epidemiological data finding that this level of activity generates measurable health benefits relative to no exercise. Yet, the recommended two-and-a-half hours of moderate activity *per week* is not nearly high enough for optimal health.[6] For that, we are best guided by the level of physical activity among indigenous people. It seems reasonable to use less active populations such as the !Kung as a reference point even if these people inhabit a desert and have relatively low food availability.

Given our high food availability, if we wanted to prevent obesity, we should at least aspire to something like the activity level of the !Kung. Of course, that is asking people to do much more than they want. Yet it is realistic given that the bodies we own were shaped by natural selection for an active lifestyle and those physiological mechanisms are highly conservative in the evolutionary sense: even experiments on rats yield realistic insights into human obesity.

We inherit a physiology that was suited for the active life of a forager but does less well in a sedentary way of life, particularly where high-energy snack foods are constantly available. Hence the epidemic of obesity and related diseases—heart disease, stroke, diabetes, cancer, joint diseases, kidney failure, and so on, that are brought on or aggravated by being overweight or obese.[7]

Such conditions constitute the biggest cause of illness in developed countries, affecting large numbers of people over long time periods.

Many of these ailments are chronic and expensive to treat so that they make huge holes in health-care budgets. stretching our capacity to cope with other kinds of diseases. This serious modern health issue represents the mismatch between our inherited physiology—that was designed to conserve food energy and store fat when possible—and modern conditions. There is no good physiological solution to this problem, although the health industry makes a fortune offering drugs and surgeries that cope with obesity and its pathological side effects. Yet the only satisfactory solution is behavioral. We need a more active lifestyle that is more in tune with what our bodies were designed to handle.

How adaptable are humans to ecological transitions such as settled agriculture and industrialization? The prevalence of obesity is sometimes interpreted as clear evidence that humans do not adapt well to changing modern conditions. Yet that conclusion is premature. We need to give ourselves a reasonable amount of time to adjust, and it can take more than a generation for old habits to die and new, better adapted ones to take their place. That point emerges from the controversy among evolutionary psychologists over addiction to tobacco.

ADDICTION TO TOBACCO AND ADAPTATION

People in many societies use stimulant drugs to help them feel energetic and capable of facing the problems of daily life. Examples include betel nuts and coca leaves that are habitually chewed in developing countries for the stimulants they contain. In most developed countries, caffeine is acquired from tea and coffee for similar purposes. Tobacco is smoked because nicotine from the smoke activates specialized receptors in the brain that produce a pleasant increase in arousal.

The problem with tobacco, whether chewed or smoked, is that it is highly addictive. Not only that, heavy smokers can expect to lose two decades from their lives thanks to many complications ranging from emphysema and heart disease to lung cancer. As if that were not bad enough, their quality of life is permanently impaired long before serious diseases emerge: they experience breathlessness and chronic bronchitis associated with smoker's cough.

Why would people hobble their own health in this way? Whereas smokers may effectively choose the addiction by choosing to use the drug initially, they develop health problems not because they *choose* to smoke but because they are nicotine addicts. Why do people smoke and become addicted? One plausible answer is that the beneficial effects are immediate whereas the adverse effects are delayed in time and therefore get discounted (if they are known).

According to some scholars, including Donald Symons of the University of California at Santa Barbara, the fact that people voluntarily engage in such damaging habits as smoking means that looking for adaptive explanations for human behavior in modern societies is a fool's errand.[8] One might point to many other examples of maladaptive behavior as well. The widespread use of contraception with falling birth rates—and declining populations—in developed countries is the most striking example. Contraception use illustrates modern humans doing exactly the opposite of maximizing reproduction. This counteracts the second and most important imperative of living creatures, namely, to survive and to reproduce. Yet the availability of contraception is not the principal reason that fertility is declining. The key problem is that fertility falls due to the high cost of raising children in modern economies. This is a problem that cannot fix itself. It calls for government intervention in some form if nations are not to get sucked into the black hole of demographic winter.

The smoking problem may not be so intractable to an adaptationist approach as Symons thinks. That might indeed seem to be the case if one restricted one's focus to a single generation when smoking rates are on the rise. People acquire contradictory information about tobacco over time, however.[9] At first, it is a pleasurable stimulant. The positive connotations of happiness, success, and the good life were depicted in the movies of the 1950s in the United States featuring stars such as Humphrey Bogart and Marlon Brando. During that period, tobacco companies conspired to bury scientific research establishing the strong link between cigarette smoking and lung cancer. Once the scientific evidence became widely known, there was a steep fall in smoking rates that continues to this day.

In the face of sharply declining sales in the developed world, tobacco companies redoubled their marketing efforts in developing countries where smoking rates and sales are currently rising. Why are residents of poorer countries risking their health by adopting smoking that is not just dangerous but highly addictive? Evidently, they still associate smoking with success and the good life. Past smoking rates were low so that they have very little personal experience of the adverse health effects of tobacco addiction. Although the relevant scientific information is now widely available, poorly educated people often discount such scientific information as irrelevant to them.

It is only when the science is borne out by personal experience that residents of poor countries take it seriously. This often means the loss of a parent, family member, or friend to smoking-related lung cancer or other illnesses including emphysema and heart disease. So adapting to the adverse health effects of tobacco addiction takes several generations, and this process is likely to be more drawn out in poor countries where residents have minimal understanding of the relevant scientific evidence. For these reasons, cigarette sales to developing nations can rise as tobacco companies replicate their diabolical attack on human health in the developed world. In developed countries, sales are declining sharply thanks to greater awareness of the scientific link between smoking and cancer. The adverse effects of cigarettes are both communicated in anti-smoking health messages and reinforced by personal contact with victims of smoking-related diseases. [10]

Contrary to Symons's assertion that human behavior in modern societies is so hopelessly maladaptive that it invalidates any analysis in terms of evolution, a more nuanced conclusion is appropriate, recalling so-called sick societies. For instance, the collapse of boat-building skills by Tasmanian Islanders was not due to an inherent weakness in the society itself, so much as the natural consequence of the community becoming geographically isolated and therefore smaller. [11] It is simply too difficult for small populations to maintain complex skills whether one studies song complexity in birds or Tasmanian boat-building expertise. Why exactly small populations lose complexity is not entirely clear but there are two good leads. First, there is a limited talent pool whether for star singers or gifted craftsmen. Second, less occupational specialization is possible so that families have less incentive to hone and instruct children in specialized skills that may be passed down to future generations. Far from being sick or disordered, the loss of boat building was an adaptive response by Tasmanians to the altered circumstances created by the loss of the land bridge originally connecting their isthmus to the mainland.

Similarly, health-damaging tobacco addiction is far less problematic to an adaptationist perspective than might be imagined. New users experience nicotine as a mildly pleasant stimulant. So the widespread adoption of tobacco by indigenous people when they are exposed to it by foreigners is not surprising. Nor is the ongoing increase in smoking rates in the developing world. It is only when people become highly literate that information about the long-term health effects of tobacco addiction are either known or taken seriously. In this regard, it is worth pointing out that residents of poor countries generally live their lives in the present and are little swayed by concern for the future. Once people become fully informed about the link between cigarette smoking and cancer, smoking rates decline.

So the rapid adoption of smoking following widespread availability of cigarettes early in the twentieth century entered an abrupt reversal from the 1970s on. [12] The decline occurred despite a distressing tendency for young people to adopt smoking as a sign of adolescent rebellion from the anti-smoking message put out by parents and authority figures. Just as adoption of smoking was a multi-generation phenomenon, the decline in tobacco use is a protracted process as well. Some changes in social behavior can take several generations to run their course, whether it is the emancipation of female sexuality and careers, the assimilation of immigrants, or the adoption and subsequent rejection of smoking.

Such dramatic change occurs by nongenetic means over a relatively short period of time compared to gene selection, which is best observed on a time scale of millions of years, as preserved by the fossil record. Changing social behavior is much faster—as illustrated by the rise and decline of smoking within living memory.

Social scientists like to analyze these changes in terms of intervening variables, such as attitudes to smoking, but this approach is flawed: the cultural determinist assumption that there are socially transmissible attitudes that determine behavior is scientifically empty. It is circular reasoning wherein whatever change occurs is seen as essentially causing itself. In other words, the use of attitudes as an explanatory vehicle is logically equivalent to the ether theory of light transmission that is similarly devoid of scientific content or usefulness. The principle of parsimony dictates that it should be omitted. Instead of saying that anti-smoking messages affect attitudes to smoking, it is better to address behavior changes and how they are produced at an individual level.

We are, of course, subject to social influences, but these are an individual matter. Children who are forced to sit through informational videos about lung cancers will see repelling images of disease that make smoking seem very unappealing. This is a visceral response, and it must be individual and vivid if it is to prevent smoking. Conversely, merely being aware of the abstract scientific fact that smokers have worse health is less likely to prevent children from smoking. Merely knowing that their parents' generation disapproves of smoking won't cut it. One cannot really understand social change without knowing how it is effected in the life of the individual. That is the only causal chain that matters and there is no point in dressing it up in the ethereal fog of cultural determinism. What do we know about the individual basis of societal changes? In practice, how do individuals adapt to changing social conditions?

PSYCHOLOGICAL ADAPTATIONS IN POST-MALTHUSIAN SOCIETIES

The big changes in human history—the Agricultural and Industrial Revolutions—are complex in their causes. Yet, in each case, we see humans being pushed into ways of life that are their best, or only, viable option. Early farmers likely became committed to agriculture when foraging was no longer viable as a result of increasing population density due to hobby farming. The improved efficiency of European agriculture from the medieval period on liberated many peasants from feudal estates and concentrated population in towns. This set the stage for the Industrial Revolution by freeing up labor and increasing trade and capital accumulation.

A big theme of this book is that people do not simply change their activities in adapting to altered environments. They get changed themselves: an indigenous hunter-gatherer is very different from a farmer who, in turn, is very different from a modern urban resident. These changes are psychological as well as physiological.

One of the key precipitating variables of the Industrial Revolution was a sharp increase in (intrinsic) work motivation. We became much more productive not just because the technology of production improved, but because we wanted to work harder and produce more (thanks to better nutrition and health, enhanced male competition over women, and a declining role for extended families whether as a sponge for absorbing excess resources or as an insurance system for workers).

Work motivation is more important as an impetus to economic growth than economists acknowledge. Statistically, it explains more than half the international variation in economic productivity. That is much more than any of the regularly mooted explanations for the Industrial Revolution from infrastructure and technology to trading networks and favorable ecology (such as that of Europe). In addition to working harder and being more intrinsically motivated, members of modern societies differ in many ways from their forebears in agricultural societies, just as farmers differed from foragers. These differences represent human adaptation to altered living conditions.

Farmers became more conformist and less individualistic as they toed the family line to ensure land inheritance and conformed to agricultural practices and annual routines that were essential for success in growing crops reliably.[13] Because the agricultural way of life was riddled with uncertainties of weather, animal diseases, and pests, farmers were more intensely religious than their forager forebears.[14] Many of the most fanatical religions of the current era exist in agricultural societies, but rising urbanization correlates with a decline in religion.

Major ecological transitions produce complex psychological adjustments, and the Industrial Revolution is no exception. These changes are pervasive, ranging from prevalent obesity and rising intelligence, to declining religiosity associated with improved health and life expectancy. These changes are well-known. This book highlights some new findings about changes in work motivation that are not merely a product of the Industrial Revolution but seem to be one of its major causes.

Although we are accustomed to thinking of the Industrial Revolution as a product of technology, infrastructure, trade, capital accumulation, and natural resources, it is becoming clear that the primary engine of development is motivation. Without changes in work motivation, an Industrial Revolution could never happen as many industrialists discovered when they moved their plants to low-cost poorer countries but could not make them profitable due to low worker productivity among other problems (including poor infrastructure, lack of education, and corrupt government).

So what did change? Why did workers suddenly become so much more motivated and productive? That is a pivotal question that cries out for explanation.

HARD WORK AND MOTIVATION

The key reasons that people work harder are fairly obvious. They must be well nourished, in good health, and motivated to work toward a better future. In "peasant" societies, farmers were obliged to work hard to survive but, once their survival needs were covered, saw little point in working ever harder to accumulate capital because social mobility was not really possible. The underlying reasons were complex. To begin with the social structure was made up of extended families that operated as three-generational subsistence units.

Extended families sap work motivation in two ways. First, any advantages of increased production are distributed throughout the extended family so that the payoff to individual workers is small, as Chayanov found for Russian agricultural cooperatives where workers limited their objectives to satisfying basic subsistence needs.[15] Second, loafers are well taken care of once they do enough to ensure an adequate food supply and meet other basic needs. As families get smaller, the benefits of hard work go to the worker and his or her offspring. Moreover, surplus earnings can be accumulated to improve living standards and social status.

The collapse of extended families boosted work motivation because each individual acquired greater responsibility for his or her own welfare, and hard work was rewarded by improved living conditions and rising social status. This change was reinforced by better health and nutrition. Improved nutrition meant that workers were more energetic and enjoyed spare capacity: they could cultivate rewarding leisure-time pursuits in the hours after work. A dramatic increase in overall health and life expectancy meant that workers became more future-oriented. They delayed grat-ification by saving their earnings to finance a more prosperous future, whether in continued gainful employment or in retirement.

One of the distinguishing features of modern societies is that workers look ahead to a long, prosperous future. In Malthusian societies, workers paid little attention to their futures, acting as though there was only the present. This wasn't an unreasonable perspective given the shortness and uncertainty of most human life. They discounted the future and were not interested in working harder today to earn a better life later on.

In Malthusian societies, it was difficult to motivate workers to go beyond the effort required to satisfy immediate needs. Drawing on anthropological evidence, one exception here was men working hard to purchase a desirable spouse. This commitment is illustrated by the bibli-cal story of Jacob working for seven years as an indentured servant for Laban to earn Rachel as his bride (only to be tricked into marrying the wrong daughter and having to work seven more years).

Outside of slavery, such long-term work commitments were virtually unknown. (One exception here is the case of subsistence societies that operate a patronage system, such as an apprentice farmer working with-out pay to be rewarded later with land ownership.) Sexual access to an attractive woman is the most common source of enhanced work motiva-tion for men in subsistence societies. This does not work in reverse be-cause women are generally more in demand as sex partners than men are. In polygamous societies where men can accumulate resources by their own efforts, and thereby obtain multiple wives, they are trained to work hard according to research by anthropologist Bobbi Low on child training practices in pre-industrial societies.[16] Even in countries where there is no polygamy—such as contemporary China—bride price increases along with the scarcity of women. Men must work harder to accumulate the bride price if they want to marry.[17]

Sex ratios change progressively in tandem with economic develop-ment because males benefit more than females from improved survival in infancy and childhood. The supply of males increases as countries devel-op. The sex ratio emerged as a strong predictor of economic productivity across countries, although it proved less reliable as a predictor of creative productivity (as measured by patent applications and book publishing).[18]

Why are countries more productive if there is a higher proportion of men than women? We can dismiss the possibility that men are just more productive than women because there is no evidence of a reliable gender difference in work productivity. In the past, whenever there was a scar-city of marriageable women, men competed more strongly for brides and this favored men with inherited wealth. Men who lacked substantial inherited wealth had to work harder to accumulate assets and rise in social status if they wanted to succeed in marrying.

In contemporary developed countries, the relative scarcity of women of marrying age is less significant due to the rise in premarital sexuality. This means that unmarried men can benefit from a large pool of sexually active single women and enjoy a satisfying sex life without being married. Under such conditions, masculine work motivation falls as the marriage market becomes less favorable to women. This phenomenon is clearly visible in the United States where female enrollment in colleges and universities now substantially exceeds that of men. This suggests that males are now less ambitious than females. [19]

So the old rules of marriage markets are subtly different. Women still look for high-earning men, whether as spouses or as romantic partners, but find that there are fewer of them around. [20] There is also a paucity of ambitious young men who work hard to improve their economic circumstances because they do not need financial success to have a satisfactory sex life. This translates into older age at marriage, lower marriage rates, and higher divorce rates. Higher divorce rates are probably a symptom of declining emotional commitment by men. [21] Women find low-commitment marriages emotionally unsatisfying and therefore file for divorce (two-thirds of divorces being filed by women in the United States). [22] Divorce rates are higher in developed countries because women in unhappy marriages have more economic opportunities and therefore greater capacity to fend for themselves.

Women in developed countries are more intrinsically motivated as workers because they look forward to lifelong careers that were denied to their mothers by early marriage and prolonged childbearing. This change is one conspicuous aspect of a broad pattern of declining gender specialization and increasing gender equality.

THE EMERGENCE OF GENDER EQUALITY

An age of machines devalues manual labor in general and the greater muscular strength of men in particular. As recently as the 1950s in the United States, unskilled men could earn enough by the sweat of their brows to rise in social status and make themselves attractive as prospective husbands. [23] Industrial jobs such as meatpacking and car assembly still had a heavy labor component that made them better suited to men. At the same time, there was a huge disparity in pay according to gender. Given their primary role in raising children, women were treated as temporary workers—and paid as such—in addition to having minimal legal and political rights. According to political scientists Ronald Inglehart and Christian Welzel,

> The rise of gender equality is another aspect of the process of human development. That is comparable in importance to the global trend towards democracy and closely linked with it. Since the dawn of history, women have had an inferior social position in virtually every society. The role of women was largely limited to the functions of reproduction and caretaking: public decision making and political power were predominantly male domains. Even today, men still dominate most areas of economic and public life [references omitted].[24]

In ancestral societies, authority was hierarchical and predicated on the use of physical force. Women were at a disadvantage to put it mildly. One thinks of Yanomamo women being abducted in warfare and living out their lives in reproductive servitude to their captors—a phenomenon that was not all that rare in pre-industrial societies. Even so, the implied denigration of reproduction by Inglehart and Welzel is itself a modern point of view. From a Darwinian perspective, reproduction is all that really matters. So women devoted themselves to their most vital domestic concern—their children—whereas men were obliged to take up the slack in other aspects of family life, including provisioning, economic activity, defense/security, and political alliances—activities that were mostly conducted at a distance from home. From that viewpoint, women manipulated men into doing everything that was necessary to raise children successfully.

Such debates are semantic. They depend on whether power is defined as getting what we want or being socially dominant. A more serious academic controversy involves how best to explain the current shift toward gender equality. True to their cultural-determinist views, Inglehart and Welzel attribute the change to rising self-expression values that increase tolerance of human diversity. As I have argued throughout this book, their approach is fundamentally flawed. It is premised on circular reasoning where the snake of rising self-expression is found swallowing its own tail of tolerance of diversity. The phenomenon being explained—increasing tolerance—is presented as its own cause. I want to propose a very different explanation based upon adaptive change.

In the early industrial world, men had an advantage in terms of earning capacity for both economic and social reasons so that prevailing conditions favored a gender role specialization, with female child care and male earning. This gender specialization was not much different from that of earlier subsistence societies (except that female foragers brought home the bulk of the food as well as doing most of the child care). This gender specialization had changed quite markedly by the 1960s in America thanks to the emergence of the service economy and several related phenomena:

- Male earnings from unskilled labor declined.
- Job opportunities favored women in the new service economy that commercialized previously free work performed by women. Manual dexterity and interpersonal skills were favored over physical strength and three-dimensional spatial ability that had been at a premium in many semi-skilled factory jobs. Labor participation by married women increased steadily.
- The cost of raising children increased due to inflation in housing, education, and health-care expenses, necessitating two incomes to maintain the living standard for a middle-class home.
- Women entered higher education and careers in ever larger numbers.
- Sex before marriage went from very low numbers of women to a substantial majority with widespread use of contraceptive pills.
- Women became much more competitive in spheres such as contact sports, and their risk profile increased as reflected in illicit drug use and driving accident rates.
- Given their increased economic importance and personal assertiveness, more women entered politics and campaigned for gender equality.

Putting all of these pieces together, it was inevitable that gender equality laws would have surfaced. Simply stated, if women are making strides in all other professions, why not in politics? With more women in politics, women's issues of fairness and equality in the workplace and other areas of social life come to the fore. Women agitated first for female suffrage, but even getting women the vote was a long struggle because it conflicted with older ideas of gender specialization. Gender equality got a shot in the arm as women became more active in other spheres of life including the entry of more women into businesses and professions during the economic boom of the 1920s. (In the United States, female suffrage was enacted into law in 1920.)

Much of this progress in gender equality was wiped out during the economic distress of the Great Depression when fewer women succeeded in finding work.[25] It was not until the 1960s, when the service economy brought on vast numbers of jobs for which women were better suited than men—given their superior interpersonal skills and manual dexterity—that gender equality seemed feasible. Despite the demand for female labor, women earned less than men even when doing the same work.

Such unfairness was tolerated in an age when women were considered temporary workers but could not be tolerated in a period when women's earnings were vital for maintaining an acceptable standard of living in families, particularly with stiff inflation in the cost of raising children. Gender equality was a bread-and-butter issue that affected the majority of households. In a world where most women are economically active, it is almost impossible to imagine that the issue of equal rights and fairness in the workforce would not be addressed by the political system because these issues are so critical for the well-being of most children and families, and therefore for the society as a whole.

Gender equality is arguably the most profound social change of our era. While wage differentials favoring men persist, they are much smaller than in the past and are partly due to the greater willingness of males to move in search of better-paid employment when women are more reluctant to abandon their established social networks.[26] Men are also more willing to work in dangerous occupations, such as coal mining and fishing, that pay well but exact high mortality and health costs in terms of accidents and industrial illnesses.

Gender equality was accelerated by the service economy, but the digital revolution is more neutral. More men than women are attracted to high-tech careers, although this may reflect a work climate that is less welcoming to women. Indeed, women are more active than men in electronic social networks that are currently more a feature of leisure than of work, something that may change in the future.[27]

Either way, gender equality is an inevitable consequence of economic development, as Inglehart and Welzel acknowledge. Of course, they embrace a circular cultural-determinist interpretation of this change rather than accepting that women's job opportunities are better in a world where mechanization neutralizes the advantage of muscular strength that had favored men in all earlier societies. The decline of gender specialization at work means that women participate more in the monetary economy and acquire greater spending power and control over property.

With economic power comes political representation and other social changes promoting the interests of women. This is a process of adaptation to new economic realities. Unlike genetic adaptation that is very slow and incremental, nongenetic adaptation is typically rapid. Exactly how much time does it take for people to adapt to new social environments like the Industrial Revolution?

THE TIME SCALE OF ADAPTATION

The time scale of evolutionary adaptation is a vexed question. We know that it took tens of millions of years for the leg bones of horses to acquire their modern form and this example is widely accepted as a prototype of evolutionary change. So evolutionary change is too slow to be perceptible in modern life. Its geological pace is preserved in the fossil record. Yet that pace is far too slow to accommodate the real adaptability of organisms and their behavior that is a hallmark of all living creatures, beginning with the capacity to respond to stimulation that is a defining feature of life. Humans are no different in their capacity to respond quickly and adaptively to environmental changes.

Human adaptations such as lactose tolerance in adults have emerged since the Agricultural Revolution—a mere blink of the eye in evolutionary time. Moreover, natural selection on beak size among Darwin's ground finches occurs in a single season. Admittedly, the proportion of various bill types in the population probably shuttles back and forth quite violently over short time periods depending on rainfall.[28] If we could examine the population over a long enough time period, perhaps the proportion of stout-billed and slender-billed finches is actually quite stable. If nothing else, this example reminds us that where there is high mortality, gene selection can occur very rapidly. Of course, that does not apply to modern humans who have exceptionally low mortality compared to all earlier populations.

Whatever the pace of Darwinian selection, societal changes are remarkably fast. One striking example from my own personal history is growing up in Ireland, then a deeply religious country where religious belief and attendance at Sunday services were among the highest in the world. At the time Ireland was a poor country wherein most of the population made a living from agriculture. All that changed thanks to a period of extremely rapid growth—the Celtic Tiger—from the 1980s on. This expansion followed the relocation of large multinational companies to the country in industries such as electronics and pharmaceuticals. Within a generation, Ireland was catapulted from being one of the poorest countries in Europe to being one of the wealthiest. True to the prediction from the existential security hypothesis, the rise of prosperity brought a decline of religion with greatly reduced religious attendance.[29]

Ireland is unusual only in the rapidity of religious change concordant with the velocity of the Celtic Tiger phenomenon. Similar changes are evident in other countries as they become more affluent. The phenomenon was extensively documented using the World Values Survey and Eurobarometer surveys over the past few decades.

According to Inglehart and Welzel:

Cohort analysis and intergenerational comparisons indicate that we are witnessing a gradual process of intergenerational value change linked with socioeconomic development, reflecting the fact that increasingly favorable existential conditions tend to make people less dependent on religion and lead them to place increasing emphasis on self-expression. These findings reinforce the evidence which . . . demonstrated that the publics of rich societies are more likely to emphasize secular-rational values and self-expression values than are the publics of low-income societies. In addition, the findings converge with the time-series evidence concerning changes in postmaterialist values . . . and the time-series evidence concerning many other variables. . . . A huge body of evidence, using three different approaches—comparisons of rich and poor countries, generational comparisons, and time-series evidence from the past two decades—all points to the conclusion that major cultural changes are occurring, and they reflect a process of intergenerational change, linked with rising levels of existential security.[30]

This can be stated more elegantly in adaptationist terms. As societies become more affluent, there is less need of the emotional support provided by religion because people feel more secure in their health and future. Given that basic survival is of less concern, they look for emotional satisfactions in their jobs and in their leisure activities ("postmaterialist values"). So intrinsic motivation rises, and creativity flourishes. These phenomena are so predictable that the level of economic development can be used to predict religiosity, creativity, work motivation, and much else besides. These changes are adaptive in the sense that they help us to make the most of our opportunities in varied environments. The underlying psychological mechanisms are by now fairly well understood.

Interjecting a layer of "values" as in the assertion that economic development strengthens secular-rational values is a needless obfuscation. What is really happening is that religion declines in affluent societies because individuals feel better about their quality of life and no longer need religion as a security blanket. Inglehart and Welzel feel compelled to mention values at every turn because that, after all, is what the World Values Survey sets out to measure. Yet values are a meaningless abstraction. Mostly without evidence, values are presumed to intervene between environmental change and behavior. In reality, behavioral change occurs as an adaptive response to environmental change, and there is no need for intervening explanations such as "culture" or "values."

Social scientists like to talk about values because they are assumed to be socially transmissible (analogous to the biological transmission of genes) and to control behavior. Values are the luminiferous ether of the social sciences, inhabiting a social world that is presumed to exist independently of the natural world. This follows the tradition of early sociologist Emile Durkheim, who referred to "social facts" as operating independently of the physical world according to a mysterious logic all their own.[31] Values certainly do not exist in the material world, and they do not change behavior any more than ether moves light. Instead, they serve to fill out an explanatory vacuum in cultural determinism. Unfortunately, it may be centuries before social scientists wake up to this fact. Like the ether, values are a conveniently seductive fiction. Their primary advantage to social scientists is that they are easily measured in surveys whereas objective behavior can be difficult or impossible to tap directly, as illustrated by the difficulty of measuring sexual behavior conducted in private.

Values are assumed to intervene between environmental changes—such as economic development—and the relevant changes in behavior. So, for example, the increased affluence in modern societies is presumed to change attitudes to religion such that atheism rises and church attendance declines. People in developed countries also become more interested in self-expression and are said to have post-materialist values. In these examples—and all others—referring to values (or attitudes) is redundant. It adds nothing to the explanation that we did not already know from observing the match between changing levels of affluence and changing behavior. We are far better off seeing this as adaptive change that fits squarely within the natural sciences like all other behavioral adaptation in ourselves and other species.

When values, and cultural determinism more generally, are stripped out of Inglehart and Welzel's theory of modernization, we are left with a theory of adaptation. So atheism is an adaptive response to living in an affluent society where wealth is fairly evenly distributed and the basic survival needs of all residents are well met.

The case of smoking suggests that adaptation to changing social environments is complex and can take several generations whether for tobacco consumption to increase initially or then to decline (as health-related problems became salient). Can adaptive social change be faster?

Inglehart and Welzel found that survival needs increase as the economy sputters (i.e., as economic activity is choked by high inflation, as happened in the 1980s) and it becomes difficult to make a living. Self-expression needs do the opposite—falling during an economic downturn. Such changes are immediate rather than generational and are striking because they may go against long-run trends associated with economic development. In other words, we see responses by individuals to twists and turns of the economic cycle. (Presumably, anyone who looked for objective confirmation of these changes would find that behavior changed in the same direction as attitudes.)

When inflation rears its ugly head and the value of wages falls, survival needs become more salient and self-expression needs get put on the back burner. Of course, this is a microcosm of the long-run changes in a country as it becomes more affluent, a twist in the road that does not prevent a nation from arriving where it is headed thanks to an overall improvement in living conditions over time despite periodic bouts of inflation and occasional recessions that follow. The key idea in this book is that people adapt quickly to changing conditions by mostly nongenetic means. Changing environments change people (and other complex animals) in predictable, immediate, and adaptive ways.

CHANGING ENVIRONMENTS CHANGE PEOPLE

Humans are transformed by changes in the environment just as other species are. Some changes are obvious and overt, like rising stature in modern societies due to improved nutrition and health or increased obesity in modern populations. Other changes are revealed only by specialized research. These include changes in personality ranging from intelligence, impulse control, and narcissism, to conformity and work motivation. There are also profound changes in social behavior, from the rising status of women to increased interest in creative endeavors, such as publishing books or applying for patents. Many of the changes described in this book are adaptive: they help people to fit in better in modern societies so that they succeed in making a living (but generally do not produce more offspring). Being able to read in advanced societies is just as important as knowing how to hunt, or gather, in ancestral communities.

Although people change to fit in with varied modern ways of life, these changes are not driven by gene selection. So what are the principal engines of change? No one disputes the power and elegance of Darwin's theory of evolution by natural selection to account for much adaptive variety in the natural world. As far as modern humans are concerned, however, Darwinian evolution does not work well for several reasons. First, and most obvious, the changes are too fast to be explained by natural selection that typically requires many generations to work.

There are a handful of intriguing exceptions. Adult tolerance of milk protein (lactose) evolved via gene selection in societies that raised domestic cattle for milk. Similarly, sickle cell anemia was favored as part of a defense against malaria in places where the clearing of trees left standing pools of water in which mosquitoes bred.[32] Increasing height might plausibly have a genetic component.

SEXUAL SELECTION FOR HEIGHT

Researchers investigated a role for gene selection of stature.[33] People became substantially taller since the Industrial Revolution thanks mainly to improving nutrition. Among developed countries, average height is greatest in the Netherlands. Reduced stressfulness of childhood may also matter (due, for example, to criminalization of corporal punishment and child abuse). This is because stress *impairs* growth—a phenomenon known as psychosocial dwarfism.[34] Environmental explanations of changing stature do not rule out sexual selection for height.

Male height is sexually attractive to women in Holland, and tall men there produce significantly more children. Because tall men have taller children, on average, the average height of the population could be getting boosted by sexual selection (a type of evolution described by Darwin where males of the species evolved traits that "charmed" the females). This might explain why Dutch men—who were relatively short 150 years ago compared to other Europeans—subsequently gained eight inches compared to a gain of only three inches for American men over the same time period.

Interesting as this study is, it is rather poor at accounting for worldwide changes in height with economic development. These are far better explained by environmental influences, specifically improved nutrition but also reduced stress from corporal punishment, exposure to growth hormone in milk and meat, and possibly more time spent under artificial light that speeds growth and development in experiments on rodents. Widespread use of antibiotics in childhood might also be a factor considering that farmers routinely include antibiotics in animal foods in order to promote growth of domestic animals. The general idea is that fighting disease requires metabolic energy and inhibits growth.

The sexual selection explanation for Dutch height raises more questions than it answers. Do tall Dutch men have more children because they are more likely to divorce and remarry? If sexual selection made Dutch men taller over the past 150 years, why were they so short to begin with? Men in the United States used to be the tallest in the world, but that is no longer true. Why? As far as this country is concerned, environmental influences can explain most or all of the increases in stature over the past century-and-a-half.

So gene selection is hardly a promising explanation for rising stature since the Industrial Revolution. Even in Holland, sexual selection accounts for only about *a quarter of an inch* of increased masculine height.[35] That is what might be expected of gradual Darwinian change. Indeed, it is actually quite rapid from that perspective. If maintained, it would amount to ten inches in a mere six thousand years. This is the blink of an eye in evolutionary time. Of course, it was not maintained in the recent past and probably will not be in the future either.

None of this discredits Darwinian selection in human affairs. Indeed, there are numerous examples of genetic evolution occurring among humans since the Agricultural Revolution.[36] Many of these are responses to living in association with domestic animals, from lactose tolerance among adults who use cow's milk as food to partial immunity against animal diseases, such as cow pox (used in vaccines against smallpox). Domestic animals exposed us to parasites such as fleas and ticks that brought many new diseases to which immunity evolved. The same is true of skin and respiratory diseases contracted directly from animals, such as ringworm and tuberculosis from cattle. Complex and varied as these rapid evolutionary changes undoubtedly were, there is little evidence of genetically mediated change since the Industrial Revolution. That makes evidence of gene selection for height in Holland quite intriguing.

We cannot rule out genetic explanations for many of the changes in human behavior and physiology concurrent with economic development. Yet gene selection cannot be very consequential given the small amount of time that has elapsed. Exceptionally low mortality also renders natural selection unusually weak. Even if conventional Darwinism is ruled out, we cannot ignore genetics because gene expression is altered by the environment—producing epigenetic change.

EPIGENETIC CHANGES

When gene expression is altered by the environment, there can be marked behavioral consequences. This phenomenon is highlighted by experiments on rodents raised in a stressful environment. These animals propagated greater fearfulness into subsequent generations, not by changing gene frequencies but by altering gene expression.[37] Instead of being a source of consistency in varied environments, genes that are expressed in one environment may be silenced in another. Notably, the altered pattern of gene silencing evoked by a particular environment can be propagated to offspring via inherited banding patterns of the genome.

Children raised in stressful homes are significantly shorter in stature despite the fact that height is one of the most genetically heritable traits. Psychological stress inhibits normal growth and development by changing the way that genes are expressed. So a psychological (i.e., environmental) influence alters gene expression thereby shortening stature. Psychosocial dwarfism is apparently epigenetic. The same is likely true of the Flynn effect of IQ scores increasing with economic development.

One smoking gun here is the fact that children who witness violence as they grow up score lower on IQ tests, implying that severe stressors inhibit the development of cognitive ability, presumably by altering gene expression.[38] Both low IQ and psychosocial dwarfism point to varied ways that environmental stressors may affect gene expression. Children in developed countries are legally protected from stressful upbringing, including corporal punishment, that could play a role in rising stature in addition to increasing IQ scores.

There are many other possible cases of epigenetic effects in child development. For instance, some specialists in the development of auto-immune disorders, such as asthma, believe that these are related to the sterile conditions under which modern children are brought up. If so, then children in developed countries are more vulnerable compared to children in less-developed countries where there is more exposure to diseases and parasites.[39] Their developing immune systems do not receive enough of a challenge from pathogens and therefore develop abnormally. Conse- quently, they experience unnecessary reactions to nonthreatening stimuli such as heavy exercise (that may precipitate asthmatic attacks). It is diffi- cult to test this theory out without experimentally exposing youngsters to potentially dangerous diseases, but it does help explain why asthma and allergies become more common in developed countries where children are raised in very clean homes that insulate them from exposure to dis- eases and parasites. Their immune systems have minimal opportunities to learn which organisms are a genuine threat and which may be safely tolerated.

Another likely example of epigenesis concerns the rise of shortsight- edness in developed countries. In Malaysia, many children wear glasses due to myopia. This high level of contemporary vision deficiencies can only be environmental given that gene frequencies in the population could not have changed quickly enough to account for such a rapid in- crease in myopia. The best available explanation is that children spend a lot of time looking at books or screens from close distances.[40] Apparent- ly, this abnormal amount of close-vision focus in childhood alters the way that the eyeball grows, making it more elongated and increasing the chance of becoming short-sighted. The strongest evidence supporting this theory concerns indigenous societies, such as the Inuit, where there were neither books nor TV, and there was also a very low incidence of myopia.

Due to epigenesis a child can develop quite differently in different environments. The genotype comes with tricks up its sleeve, so to speak. Development can go in different directions, and which genes get ex- pressed is determined by environmental cues. This phenomenon surfaces in environmental effects on intelligence in different countries—influ- ences that range from parental stimulation and stress to neurotoxins.

How well a person does on cognitive tests throughout their lives is affected by early experiences that impair normal brain development. These include poor maternal nutrition in pregnancy that reduces birth weight, exposure to infectious diseases, contact with environmental neurotoxins such as lead and manganese, and reduced verbal stimulation from caregivers. The known effects of these experiences on IQ (and even on brain size) are most likely explainable in terms of altered gene expression throughout early development. What is true of intelligence also applies to work motivation and productivity. Researchers now recognize that there is a strong and clear connection between nutrition and school performance, but nutritional inadequacy also saps the work motivation of adults—they need to take more days off and cannot work for long at a high level of intensity.

According to economists Roderick Floud, Robert W. Fogel, Bernard Harris, and Sok Chul Hong:

> On the basis of all this evidence from the modern world, it seems highly likely that populations in the past, with high levels of malnutrition, suffered from low productivity not simply because of diminished physical strength but also because of diminished cognitive ability or intelligence. The relative contribution of these two potential causes of low productivity may be impossible to determine, but they are likely to have reinforced each other.[41]

Underlying this relationship between nutrition and intelligence, there is, of course, a complex interaction between nutrition—at all phases of development—and brain development. In other words, the overall level of nutrition, in addition to specific nutrients, such as fats, early in brain development affect the expression of genes that affect brain structures and anatomy. Interestingly, tall people earn more, and their increased earning capacity is attributable to higher intelligence, both phenomena being predicated on good prenatal nutrition.[42]

Although it is difficult to rule out epigenetic explanations for many of the changes described in this book, we can nevertheless be confident that conventional gene selection has little or nothing to do with the psychological changes associated with economic development. The changes are far too rapid to be plausibly explainable in terms of altered gene frequencies of the population. What is more, these alterations of intelligence, work motivation, health, childbearing, future orientation, religiosity, and so forth can be readily explained in terms of environmental influences that are at play in all developing countries. Adapting to altered environments—rather than genes or "values"—is the real engine of change, and some of these changes involve epigenetic effects.

We live at an exciting period to study social transitions because change is happening so quickly that it is difficult for scholars to make sense of it. So we are just beginning to understand the dynamics of marriage markets at a time when people in social democracies are turning their backs on marriage, for example. Encouragingly, all developing countries undergo the same kind of changes and these are best understood as nongenetic adaptation to changing environmental conditions. Economic development mirrors another major transition, namely the Agricultural Revolution.

MAJOR ECOLOGICAL TRANSITIONS

Intelligent animals benefit from learning how to do something differently. They exploit new food sources such as rats in Italy diving to the bottom of a river to take shellfish or chimpanzees fashioning a tool to fish for termites or crushing large nuts between an improvised hammer and anvil. Even seemingly less intelligent animals can make remarkable changes when conditions change. One is impressed by the capacity of the Mauritius kestrel to begin nesting on cliffs and thereby foil newly introduced predators on the island.

Behavioral adaptability of this sort is quite common, as is the ability of animals to learn useful habits by observing what others around them are doing—a propensity that is likely present in all social vertebrates. [43] The Mauritius kestrel illustrates adaptation to the environment in a single generation via social learning whereas genetic adaptation generally requires tens of generations (except in the specific case of genetic polymorphs like cannibal salamanders where one morph can take over in a single generation). Of course, the underlying mechanisms are very different in these examples. One is a case of behavioral modification whereas the other implicates adaptive development acting on bodily growth for tiger salamanders.

Humans experienced two major transitions—the Agricultural and Industrial Revolutions, both driven by environmental change. The Industrial Revolution really was a revolution in the sense that people aspired to a life of ever greater affluence unlike anything experienced before on the planet except for what happened within the castles of a monarch like Louis XIV. There was nothing "revolutionary" about agriculture: the term is misleading. To begin with, there was no sudden breakthrough in knowledge. Our ancestors understood agriculture very well and had likely practiced it as a part-time subsistence strategy for thousands of years before settling on farms. Farmers were rather poorly nourished, suffered from many new occupational disabilities like joint deformities, and experienced epidemic diseases for the first time due to increased size and density of the human population that constituted a reservoir of infection.

The fact that the population increased so rapidly following the switch to settled agriculture proves that farms produced much more food than could be collected from the natural ecology. Instead of being a "revolution," agriculture was an uneasy compromise. Imagine that a bush pilot develops vertigo and can no longer fly. He returns to his former occupation as a tax accountant. That captures the letdown that our ancestors must have experienced when forced to give up their exciting and emotionally satisfying lives in the forest for a humdrum existence characterized by unending hard labor, obedience to routines, and lack of excitement and individualism.

Although preferable to agriculture, there was nothing easy about hunting and gathering. After all, the human population was kept down to very low levels. This was partly because large tracts of land were necessary to support a small foraging group: any greater concentration and their prey animals were hunted out of existence. The reproduction rate was also low because children were breastfed for several years, which suppresses ovulation and creates large intervals between births. Even so, mortality rates must have been high from animal attacks, infected injuries, vector-borne diseases like malaria, and homicides.

The sharp rise of population following the Agricultural Revolution suggests that living conditions were getting better but the opposite is true. Like other species, humans lived in a Malthusian trap. As food production increased, the population rose until it reached the carrying capacity of the local environment. The end result was always the same—hunger, or too many mouths for the available food supply. Thomas Malthus concluded that there was no way out of this trap. Or, rather, the only solutions were the Four Horsemen of the Apocalypse—death, disease, famine, and war. These removed population so that the survivors had more food. This respite from the threat of starvation was temporary. Food scarcity would materialize again as soon as the population recovered to its carrying capacity.

In contrast to the two earlier ruts of humankind, the Industrial period really was a revolution that permitted real increases in the standard of living. Contrary to Malthus, industrialized farming produced more food than was needed—even in the face of a rapidly growing population. It did not solve the problem of global hunger, of course, but that turned out to be more a problem of food distribution rather than food production. The planet actually produces far more food than is needed, but a lot of the bounty gets wasted. Moreover, regions that experience periodic droughts and other forms of crop destruction cannot receive the surplus from elsewhere due to logistical problems and transportation expenses.

The Industrial Revolution got humans out of the Malthusian trap. Just as important, we have seen a steady and continuous improvement in the quality of life by all conventional measures. People must still die but important inroads have been made in controlling all forms of disease, in reducing violent deaths of all kinds, and in minimizing hunger. Defeating three out of four Horsemen is not bad!

Even Death is being held back to a striking degree. The most impressive achievement of the twentieth century was a doubling of life expectancy in leading countries such as the United States. This is important not just because the average person cheats death for twice as long but because it speaks volumes about improvement in living conditions.

Better quality of life is illustrated by the remarkable increase in real wages over the past two centuries. During the twentieth century alone, the standard of living increased by a factor of six or seven in developed countries by this reckoning.[44] By other real metrics (that get around the distorting effect of monetary inflation), the improvement in living conditions since prehistoric times has been even more spectacular. In one of the most entertaining of these measures, Matt Ridley points out that an hour of work in 1750 BC yielded 24 lumens of light compared to 8.4 million lumens in the present.[45] This estimate leverages improvements in technology between a rush light and a compact fluorescent bulb. Illumination technology has undergone significant further changes since Ridley wrote his book. The upshot is that artificial light today must be close to a million times cheaper in work cost than it was for the ancient Egyptians.

Overall spending power of workers also rose at an astonishing rate, yielding the impression of overwhelming increases in the standard of living, as documented in *The Changing Body*.[46] At the end of the twentieth century, European workers earned more than four times as much in real terms as they had at the beginning of the century. Even more dramatic improvements were seen in the United States where manufacturing hourly wages increased from $0.85 in 1910 to $14.32 in 2000 (in 2000 dollars). Money alone, and the material products it yields, might be meaningless if it was obtained by a life of constant toil. Yet the opposite is true. Employees today work far less than their counterparts of a century ago. For instance, Americans work 30 percent fewer hours than they did a century ago. There is a corresponding increase in the amount of leisure time—from about 1.8 hours in 1880 to 5.8 hours in 1995, for U.S. heads of households.

Improved standards of living are also reflected in altered spending patterns. The proportion of wages spent on basic necessities, particularly food, dropped from 49 percent in 1875 to 5 percent in 1995. During the same time period, spending on leisure increased from 18 percent to an astounding 68 percent. Whereas food, clothing, and shelter have all become cheaper, the cost of education and health care have risen sharply, in part because people are receiving more medical care and pursuing more advanced degrees. When one combines huge increases in leisure time with corresponding increases in leisure spending, this implies that the lifestyles of contemporary workers are similar to those of the idle rich in earlier generations with plenty of leisure time and expenditure of huge amounts of money on leisure pursuits.

Although American workers toil for fewer hours than in the past, many put off their dreams of a still more leisured lifestyle until after retirement. A century ago, few men retired, and most died in the harness, so to speak. The labor force participation of American men sixty-five and older stood at 63.1 percent in 1800 but fell to just 17.7 percent in 2000.

With earlier ending of work and increased life expectancy, retirement is becoming a much more significant proportion of people's lives. So a worker who retires early at the age of fifty-five and lives to be eighty-eight spends as much of their lives in retirement as in the workforce (thirty-three years, assuming they complete their education by twenty-two and remain employed throughout their lives). Retiring this young is admittedly unusual outside of public employees such as police, but it is becoming more common. Early retirement opens up vistas for self-improvement and self-actualization that were unheard of in earlier generations.

Considering that real hourly wages increased by a factor of fourteen between 1910 and 2000, there has been an unprecedented rise in the standard of living of ordinary people with the potential for greatly increased discretionary spending and more resources to invest in homes and save for retirement. Such improvements in the standard of living of ordinary people are in marked contrast to the Malthusian world that prevailed before the Industrial Revolution. In that grim world, the only hope for improvement was a temporary drop in population that left more resources for survivors. These additional resources were soon expended to raise more children who restored the population to its carrying capacity. This reestablished the gloomy reality of a benighted world where the quality of living never saw any sustainable improvement.

An improved standard of living for a greatly expanded global population certainly gives the lie to Malthusian pessimism, but it is not an unmixed blessing either. More affluence for more people translates into more consumption of goods and services. This is equivalent to more energy consumption and more destruction of natural resources such as rain forests and fisheries. It raises the specter of climate change and ultimately making the planet unfit for human habitation. Such problems are at least partially solvable by changing human behavior. Most viable solutions require somehow reining in human consumption, at least to the extent of using less energy.

Environmental problems are outside the scope of this book, but they do suggest that Malthusian limits are reinstated albeit at a much higher level of individual consumption and personal affluence. When the eventual decline of global population is combined with wiser global stewardship, we may still end up with a far better quality of life for ordinary people than Malthus would have considered possible in his wildest dreams. So humans could end up in a sustainably better place. I argue that the biggest reason for this change is motivational whereas historians emphasize technological innovation as the main engine of change. How can this conclusion be defended going as it does against the grain of historical scholarship?

THE HUMAN SIDE OF CHANGE

Technology certainly matters. The Industrial Revolution in England would hardly have occurred without the use of steam engines in factories. Similarly, the contemporary economy is in the process of shifting from people and buildings to apps and the Internet. According to my analysis, countries having the greatest Internet penetration are also the most productive.[47]

Technology is not unimportant. However, human motivation matters a lot more than technology in explaining why some countries are much more productive than others. This is an empirical issue, and the evidence is unambiguous. More than half of country differences in productivity are related to Darwinian competition as reflected in improved nutrition and health, the collapse of the extended family, and tighter marriage markets for men. The marriage market, as such, is far less important today than in the past thanks to a decline in the importance of marriage itself, and greater gender equality in the workplace, but the mate market for sexual partners is still in play.

Such sweeping conclusions excite skepticism, particularly when they run counter to received wisdom. How can skeptics argue with the data? If Darwinian competition accounts for differences in productivity between countries (and over time) better than technology, climate, government, and a score of other rival explanations, how can one possibly question a Darwinian interpretation of economic development? The answer is that most of the evidence is correlational and therefore open to alternative explanations. In this case, the most serious logical problem is the possibility that the order of causality is reversed. Instead of the four elements of competition raising productivity, perhaps it is economic growth that boosts each of these spurs to work motivation.

That general criticism is unsatisfactory. If one takes health and nutrition, it is very clear why better-nourished people and healthier people are more competent, motivated, and productive. Sick or starving people avoid unnecessary effort. They are concerned about their quality of life in the present and have less interest in working hard to accomplish goals in the distant future that may not materialize for them. There is far less evidence for work motivation and the marriage market. What little evidence there is shows that men work harder when there is a scarcity of marriageable women. Conversely, when large numbers of women are sexually active outside marriage, men devote themselves to direct mating effort (as opposed to working hard in careers). For college men, this means choosing to party at the expense of studying. That leaves declining family size as the final element of Darwinian competition.

How do we know that extended families inhibit work motivation? The case is largely theoretical. We do know that Russian agricultural cooperatives were minimally productive due to lack of motivation by members to do more than required for mere subsistence. A similar rationale applies to any small cooperative economy, including households.

Children who grow up in societies that emphasize extended-family responsibilities are less individualistic in their goals and therefore less interested in accumulating wealth or aspiring to a greatly improved standard of living for themselves.[48] Health researchers in the Italian American village of Roseto, Pennsylvania, also discovered that villagers were content with a much simpler lifestyle in three-generation families. When the younger generations moved out to new housing developments, they became much more materialistic, insisting upon having plenty of space and all of the amenities preferred by mainstream Americans.[49]

So, the general argument that the four spurs to work motivation boost economic growth rather than simply being its result is plausible. The more information we have about them, the clearer the causal mechanisms are. The clearer the causal mechanisms become, the more unambiguous the direction of causality. People become more productive if they are better nourished.

Clearly, developing societies get better at nourishing all of their children. More resources are available for school lunch programs and so forth. Such programs improve school performance and help these youngsters to grow up into more productive adults. This is a positive feedback loop: more affluent countries help their youngsters to achieve greater productivity in successive generations thereby raising economic production that continues to improve nutrition in a virtuous cycle at the societal level.

This book emphasizes the human aspect of the Industrial Revolution. Like other major transitions for our species, this is best interpreted in terms of adaptation to an altered environment. Focusing on four different motivational engines may make the process seem more complicated than it needs to be. For all are different facets of Darwinian competition tapping human adaptation to the modern environment.

This is a distinctly different approach from the two major intellectual streams in the social sciences. Cultural determinists claim that we are not adapted to the environment at all. Evolutionary psychologists argue that we are not adapted to the current environment as much as our ancestors prior to the Agricultural Revolution. I hope my readers are convinced by now that each of these rival approaches is mistaken—just like the vital alternatives offered by a vacuum cleaner salesman who gives the punter a choice between the expensive brand and the cheaper one. Don't buy either of them!

WE ARE ADAPTED TO THE MODERN ENVIRONMENT— BUT NOT BY GENES

Students of the social sciences are offered two rather unpalatable alternatives for explaining human adaptation to the modern environment. Evolutionary psychologists believe that human behavior is adaptive but that we are adapted to a way of life that our ancestors led from about two million years ago instead of to modern conditions. Cultural determinists believe that our behavior is entirely homocentric and divorced from the natural world making notions of adaptation irrelevant.

Neither of these approaches helps explain how modern populations adapt to changing social and ecological conditions. Of the two, evolutionary psychology is the more credible because it is, after all, a scientific theory. Cultural determinism is not. The key mistake that evolutionary psychologists made was to overemphasize genetic determinism in their explanatory framework. Other species adapt to modern environments through a variety of different mechanisms, and the great changes we see in modern life are independent of gene selection. Such recent changes are incomprehensible to evolutionary psychologists. They provide a plausible explanation of gender differences in most societies around the globe, for example, but have no explanation of gender *convergence* in the modern world. Genetic determinism rationalizes cross-cultural similarities in gender-related behavior but has no explanation for the disappearance of gender differences. If it fails to explain modern behavior, then it is likely to provide a misleading picture of ancient behavior as well. After all, we continue to live in the same world that functions according to the same scientific laws.

By pinning their hopes on genetic determinism, evolutionary psychologists pitched a theory that is clearly irrelevant to modern life, unless you happen to believe that complex human behavior is genetically determined and plays out similarly in all generations due to genetic inheritance. That approach has serious shortcomings at all steps of the argument. There are no genetic programs that determine complex human behavior, and most of the things we do are reconstructed in every generation based on our experiences.[50] Of course, the same is true of other species. Chimpanzees fish for termites in completely different ways, using different tools, in various local communities. It would be silly to claim that fishing for termites is genetically encoded in the face of this evidence. This foraging activity is clearly an emergent property of various chimpanzee attributes that do have heritable genetic components, including opposable thumbs, manual dexterity, large brains, intelligence, sociability, and so on. Being tied to genetic determinism impedes understanding of how and why chimpanzees fish for termites, yet fishing for termites is obviously an adaptive behavior in the sense that it helps chimps to survive and reproduce.

CONVERGENCE

The connections between health, nutrition, the sex ratio, extended families, and the economy help to explain why modern societies are becoming so similar. With increased productivity, Shanghai and Singapore converge with New York or Paris in terms of their economic activities, infrastructure, lifestyle, quality of life, and political or religious views. Some of the main lines of convergence were outlined at the beginning of the last chapter, but *why* are developing countries moving in the same direction regarding fertility, political freedom, gender equality, quality of life, religiosity, freedom of expression, and so forth?

If another species underwent a profound change of this magnitude, in this short a time, we would look to changes in their environment as the probable cause. The problem is that we do not have a convenient analogy in the animal world to shape our thinking. One instance of dramatic environment-induced change involves salamanders.

Tiger salamanders grow larger and develop bigger jaws when the population density rises. Such changes are rapid and profound, but they may rely on heterozygosity, or having alternate genetic potentials where environmental variation triggers one of two developmental pathways being followed.

Locusts provide an even more striking parallel: their social behavior, reproduction, and general brain biology are predictably altered by rain falling in the desert.

The human brain manifests similar flexibility where adult behavior is affected by key early stimuli including maternal affection, psychological stress, and nutrition. (Of course, the same general principle applies to all mammals.) For humans, modern environments promote higher intelligence, greater intellectual curiosity, and interest in self-expression, all of which increase with economic development. Although there is no clear genetic explanation for the changes in human psychology and behavior since the Industrial Revolution, it would be foolish to dismiss biological influences especially given the strong relationship between good nutrition and health and national productivity.

Biology affects how modern human populations adapt to the greatly changed contemporary environment. Yet we are forced to go beyond genes and epigenetics to understand how radically we are changing to fit in with the modern economy. Epigenetics certainly matters in the sense that many of the key early influences on psychological development affect gene expression. So when children are exposed to psychologically stressful rearing environments, they are at risk of reduced intelligence, impulsive violence, delinquency, and precocious sexuality.[51] These characteristics almost certainly involve profound changes in brain structure and function that are the product of altered gene expression. Yet the explanatory power of epigenetics is limited based on our current understanding. One cannot account for the Industrial Revolution in terms of altered gene expression. There is a lot more going on.

If not genes, then what? I have been at pains to dismiss "culture" as a respectable scientific explanation for any of the changes surrounding modernization of economies. Yet social learning is real, and its scientific respectability is bolstered by its effects on all social animals with backbones (i.e., vertebrates). Human social learning is illustrated very clearly by generational shifts, such as the naturalization process of immigrant populations.

The younger generations of Roseto's Italian American residents began spending more time with friends than family and acquired their interests, goals, preferences, and habits from large houses to outsize meals, for instance. This happened partly through mere exposure to a different way of life and partly due to an exaggerated desire to be accepted by peers, particularly for minorities who seem different from the mainstream. Another much-discussed example of social learning involves smoking cigarettes, which is the focus of a controversy about whether humans can adapt to their current environment as other species are known to do.

When Americans became aware of the link between cigarettes and lung cancer in the late 1960s, the smoking rate started a decline and has sunk steadily ever since, illustrating the adaptive nature of social learning (in the sense of improving health or survival prospects). Learning to smoke and then learning to avoid smoking are not so very different, in principle, from small mammals learning what to eat by observing their mothers and learning to avoid toxic foods that are avoided by the mother.

One key departure in contemporary human behavior is the importance of mass media as vehicles of learnable information. The introduction of television broadcasting made people less satisfied with the material quality of their lives. That phenomenon gets exaggerated in the modern world of continuous electronic contact with other people, including strangers, across the globe. For contemporary humans, the social network is vastly extended so we can be affected directly by what others do and say in distant countries. At present, the worldwide communications network is so new that it is difficult to predict how future populations might respond to it. Indeed, it is hard to determine if this change can greatly improve our quality of life, as the Industrial Revolution did, or propel us into a Kafkaesque nightmare like that depicted in the Netflix series *Dark Mirror*.

HOW SOCIETIES CHANGE

Social learning is real and important for our species as it is for others. Yet it is just one of the many mechanisms through which we become adapted to the very different modern world that emerged following the Industrial Revolution.

The more demanding cognitive environment of modern life induces children to grow up better equipped to handle a vast flow of information quickly—the basic mechanism underlying the Flynn effect. Such environmental enrichment effects are found for other mammals and the level of fear or stress in the early environment is a key influence on willingness to explore and learn.

When children grow up under stressful conditions characterized by material insecurity, they tend to be less ambitious and forward-looking. They try to seize whatever happiness they can squeeze out of the present. They are also more impulsive and violence-prone, traits that characterize preindustrial societies but decline markedly in societies that experience economic growth.[52] The modern environment of children is consistently different from the past. This plays an important role in the convergence of modern societies. Cognitive ability rises, violence falls, and many other changes ensue, including the increased social tolerance and gender equality of modern societies.

In addition to these multifaceted developmental phenomena, people respond predictably to what others in the society are doing. Sexual behavior changes in response to varying marriage markets, for instance. In modern societies, most women are sexually active before marriage, and those who postpone sexual intercourse until after marriage put themselves at a competitive disadvantage because men are attracted to less sexually restrictive women. Ironically, in restrictive societies, it is the sexually liberated women who are at a disadvantage, both because they are at risk of unwanted pregnancy and because their marriage prospects are destroyed pointing them in the direction of poverty and prostitution.

Although scholars are fond of describing such dynamic processes in terms of changing moral values, such cultural determinist explanations are scientifically empty. What is really happening is that sexuality and marriage have market-like characteristics. How a person behaves is very much determined by what the preponderance of other individuals are doing, as Elizabeth Cashdan concluded in her research on cads and dads in college dating. [53]

So much for the prevalence of premarital sexuality. What about the rise of gender equality that is also often attributed to changing value systems? Because values explanations are not true explanations at all, it is better to investigate objective factors such as the movement toward gender equality in wages and political life. In particular, we see far more participation by married women in the workforce today than in earlier generations—a change that is rooted in the economic reality of expensive child-rearing conditions. This shift is accompanied by a marked decline in fertility and convergence of the work lives of men and women.

Of course, these two trends are tightly linked: modern women devote far less time and effort to raising children and far more time to generating earnings. In other words, the complementarity between male earning and female child care went by the wayside as men did more direct child care than before. There are many reasons for these changes. One of the most important is the availability of effective contraceptives: women find it much easier to regulate their fertility and delay the birth of children in favor of developing careers. Another big factor at play is the rise of the service economy and much greater demand for female labor as domestic services that had been "free" in the sense of being performed by women in the home without pay were transferred into the monetary economy, whether as paid services or as mechanical devices that performed domestic work, such as laundry machines and vacuum cleaners, which were designed to reduce domestic work performed by women.

Although many service-industry jobs were, and are, poorly paid, the great demand for female labor nevertheless empowered women, giving them a more important economic role than was true of earlier generations. Not all of the service jobs were poorly paid, of course, and modern women are enrolled in higher education at a greater rate than men, implying that they compete more vigorously for better-paid careers than men do.[54] Such improvements in economic power almost inevitably lead to greater political power, which is not to deny that women still have to fight for fair pay and equal treatment in the workplace. The key point is that changes in their economic activities are being accompanied by significant increases in political power and generally improved working conditions relative to men.

Women no longer expend nearly so much personal effort in raising children: they have fewer babies, begin their families later in life, and rely more upon commercial daycare providers as well as the assistance of husbands. So they can pour a great deal more effort into businesses and careers than was possible for earlier generations of mothers. With a change in the practical realities of their lives comes a need for greater personal autonomy and greater political freedom. Needless to say, these economic changes in the lives of women inevitably alter relations between the sexes as women enjoy greater freedom in their sexual relationships that accompanies rising economic independence. Of course, gender equality is just one aspect of modernization—one element of the great convergence that makes all modern societies surprisingly similar against a backdrop of superficial diversity. That convergence obviously occurs independently of genetic change. Yet it is adaptive in the sense that it is a predictable response to changes in the economic landscape that helps people to prosper in the modern world (even if it does not boost their reproductive success).

THE GREAT CONVERGENCE IS ADAPTIVE

People improve their own lives by working hard in a growing economy thereby responding to opportunity. Economic development—like the Agricultural Revolution before it—therefore has an important motivational component that is frequently overlooked by historians and economists.

We are accustomed to thinking of the Industrial Revolution as primarily a technological event predicated on development of new power sources that drove more elaborate machines. Yet technology is not the critical driver of change. Indeed, growth in the British economy was evident a century or more prior to the 1860s—a period christened the

"Industrious Revolution"—when textile orders were "put out" to cottage weavers who used hand looms. Better lighting permitted cottage weavers to work long hours. They were strongly motivated to work harder because doing so allowed them to improve their standard of living and rise in social status. [55]

Harnessing steam power undoubtedly created the potential for much higher productivity, although this was contingent on the willingness of workers to exert themselves. Perhaps for this reason there is a surprisingly weak relationship between investment in infrastructure and economic productivity. It is difficult to quantify infrastructure spending satisfactorily because there are so many distinct categories.

One way around the problem is to lump all government spending together and assume that countries with high overall spending are likely to invest more in shared infrastructure such as roads, energy networks, public education, scientific research, and communications systems. Government spending is one key source of funding for infrastructure and technology, but it plays an insignificant role in economic productivity once Darwinian competition is taken into account. [56] Evidently, the Industrial Revolution was caused more by a change in people than a change in technology.

Specifically, work motivation increased because for the first time in history it was possible for ordinary people to achieve a sustained increase in their standard of living and quality of life. Darwinian competition favored hard work thanks to improved health and nutrition, smaller households, and increased marriage market competition among men. Every relevant metric backs up these claims whether it is the meteoric rise in real wages throughout the twentieth century or the unprecedented increases in nutrition, cognitive ability, life expectancy, leisure, or home quality. [57] Of course these positive metrics are accompanied by declines in the four horsemen: war, death, disease, and famine. Violent crime declined to historically very low levels, and most major disease epidemics were brought under control thanks to improved sanitation, clean drinking water, antiseptics, antibiotics, vaccination programs, and improved food hygiene.

The quality of life took a distinct upward tilt. That is not just a gift from many improvements in medicine, engineering, and science. Instead, it is the product of an environment that motivated hard work among ordinary employees as well as managers and creators of intellectual property that were the focus of David McClelland's work on need for achievement. [58] Why did workers suddenly began working harder? They did so because they aspired to a better standard of living. Why did workers begin looking forward to a better life? They did so because it was suddenly possible thanks to rising real wages (facilitated by increasing productivity). At the same time, workers were primed to exert themselves thanks to improving health and nutrition, to increased competition by men over women, and to the demise of extended families that penalized effort in earlier societies due to freeloading by family members. The Industrial Revolution ushered in a major transition in living conditions. For the first time in history, ordinary people who worked hard could improve their standard of living continuously (compared to temporary advances such as those following the Black Death).

With economic development and an improving standard of living, there is an increased sense of existential security. That is, we expect to lead longer, healthier lives than our ancestors and act accordingly. We become more future-oriented. In subsistence societies, people place little store in the future because they lack confidence of living to a ripe old age. That mind-set is inimical to planning for the future or taking business risks. Instead, residents of Malthusian societies live in the present and do what they can to make themselves feel good in the here and now. They use religious rituals to dull the pain and anxieties of everyday life.

Instead of building an earthquake-proof home, Malthusian populations intercede with the supernatural so as to direct the tremors elsewhere. Once people solve the key practical problems that threaten survival, and once they feel more secure about their personal futures, they turn away from religion. In the most developed countries today, the majority of the population sees religion as unimportant in their daily lives. Ironically, they become less materialistic and more interested in developing and expressing their inner lives. [59]

So the convergence of modern societies is built upon similar adaptive changes in individuals as they adjust to a very different modern way of life. We see increased intelligence, more religious skepticism, gender equality, and a great deal of tolerance for ethnic and sexual diversity that is now codified in anti-discrimination laws. All of these legislative efforts are designed to protect the rights and personal security of the individual against the coercion and injustice that was more typical of earlier societies. We see in such protections a recognition that if we wish to guard our own rights and personal security, we can do so most effectively by protecting the rights of all. In stressing the advantages of modern life in solving the intractable problems of history, it is easy to overlook changes that have taken place in ourselves as we respond to modern conditions.

ADAPTATION TO MODERN LIFE AND THE LOCUST ANALOGY

Scholars since Aristotle, and probably well before, looked to social insects, such as ants and honeybees, for a convenient analogy with human societies. In a beehive, tens of thousands of individuals behave in an orderly fashion. The various castes perform their allotted functions so that the colony works in perfect harmony. The hive is defended by soldiers. Scouts go out in search of food, and foragers exploit whatever sources of nectar and pollen are available. The queen and the drones do no work, sticking with the job of generating new workers who are fed and cleaned by workers within the hive until they can perform useful work themselves in caring for the young and maintaining the nest. Even a casual observer who knows nothing about bee specialization is struck by the order and effectiveness with which tens of thousands of individuals coordinate their activities for the benefit of the colony.

Moralists love the social insects because they illustrate key ethical principles, including the merit of hard work and saving for the future, the capacity for self-sacrifice given that soldiers perish when they sting an attacker, and, above all, the propensity for cleanliness and order. Who would not be inspired by these wonderful creatures who may well be mindless robots but nevertheless hold up a mirror to many human failings? Bees even offer a parallel to caste systems in human societies. The drones resemble the idle rich. The sterile workers are compared by evolutionists to nonreproductive groups such as Catholic priests and nuns.

The bee analogy may be beautiful, but it has one serious limitation. Bees and other social insects, like ants, continue to do the same thing over millions of years. Human societies are characteristically changing and unstable. Urban societies of the past all collapsed. Ecological instability of human civilizations seems inevitable.[60] This periodic intensification and collapse resembles locust swarms. These devastate their ecology and are forced to resume a solitary way of life. We may be better represented by locust societies than by the eternally stable bee society where overpopulation of the hive is curtailed by the formation of new colonies.

Human complex societies are quite unstable, however, rarely making it past a few thousand years. The underlying reasons are poorly understood. One common theme is that complex human societies stretch the resources of their local ecology. The earliest irrigation-based societies poisoned the land via progressive salinization, for instance.

Another recurring problem is that central administrations become increasingly inefficient. Either they accumulate debt to wage pointless wars or maintain a bloated elite class that makes unreasonable demands on workers. Both issues weakened the Roman Empire.

History reveals that human societies are nothing like bee societies. Change is the key feature, and chronic instability is the result. Such phenomena resemble the society of locusts more than that of other social insects like bees. Of course, locusts begin as solitary insects who form social aggregations when the population rises. This transition is not just superficially similar to the transition from human foraging to urban societies. In each case, the change brings profound alterations in brain biology and social and sexual behavior in addition to the more obvious increases in population density and consequent ecological stress.[61]

At first blush, locust swarms might appear to have little in common with human cities. Yet there are many functional similarities. Both fit into a pattern of temporary exploitation of abundant resources. In each case, population size increases rapidly, stretching available resources to their limit. This phenomenon played out in the earliest civilizations where irrigation over several generations would have poisoned the land.

Such hydraulic civilizations permitted a huge increase in food production and a corresponding rise in the population of towns. Hydraulic societies were a temporary aberration in the course of human adaptation to otherwise food-poor environments. Of course, this change was also permitted by an increased supply of water, due not to rain but to the building of canals and other water-control technologies, such as drainage, terracing, and construction of elevated fields in swamps.

The ecological instability of large human populations is hardly in question today. We are repeating the failed experiments of the early hydraulic civilizations. The difference is that our contemporary experiment in overconsumption is global rather than local. It is spreading around the world based on extremely cheap ocean transportation and integrated supply chains and labor markets.

Economic development is giving us a better quality of life than was possible in the static societies of the past when Malthusian principles applied. The problem, of course, is that the price involves crashing delicate global ecosystems one at a time. The threat is perceived to be so serious that future-minded business leaders, like Elon Musk, are seriously contemplating the need to place colonies on other planets.

Overuse of resources follows on from the population explosion that is itself attributable to the success of industrial agriculture at raising food production and confounding Malthus's view that food production is rigidly fixed. Rising food production today serves a similar function as rain in the desert for locusts or irrigation in ancient Mesopotamia.

Human population growth in the contemporary world does seem much slower than the emergence of locust swarms. Yet, if one considers the generation time of locusts that is only a matter of a few months, they are more comparable than one might imagine. For example, in the locust plague that ravaged north Africa in 2004, the population passed through four generations in ten months indicating a generation time of two-and-a-half months—that is somewhat shorter than the nineteen-week generation time of solitary grasshoppers—and that increased numbers by a factor of twenty.[62]

If a locust plague may unfold in a few months, the human population explosion emerged in the past two centuries. In each case we are dealing with fewer than ten generations. That sort of population acceleration is rare among mammals (being restricted to animals with population cycles, such as lemmings and snowshoe hares) and is completely without precedent among the great apes who are notably slow breeders.

Details of the emergence of locust swarms offer many parallels with the emergence of complex sociability in human urban centers. Enhanced social stimulation alters the brain and social behavior for humans as well as for locusts. Such environmental adaptations are receiving the attention of researchers, but there are also curious similarities in reproductive adjustments that receive less attention.

Perhaps the biggest myth about locusts is that they are a disorganized mob composed of independent agents that wreak havoc purely through the impact of large numbers. The truth is far more complex and interesting. Locust swarms emerge in distinct stages.[63] Gregarious phases of the grasshopper are different enough from the solitary phase to be seen as different species. This is manifested in coloration differences but more dramatically in behavioral variation. Even before maturity, the *gregaria nymphs* are attracted to each other and form bands of several thousand individuals who migrate together like a tribal horde. Their mutual attraction is olfactory as well as visual. The bands travel together probably using the sun as a navigation aid.

Marked differences in morphology, development, and reproductive physiology have been reported. These call attention to analogous, if biologically different, adaptations of human populations to urban life as revealed in developed countries. The most interesting of these changes concern the pivotal Darwinian topics of body design, reproductive physiology, and reproduction rates.

Among locusts, the swarming phase, or *gregaria*, are darker and develop highly contrasting yellow and black markings. They have slower early development, allowing them to grow larger, but become sexually mature at an earlier age. Although they reproduce earlier, their number of offspring is reduced. These reproductive changes are likely a response to increased availability of food.

If so, they provide a fascinating analogy to humans in developed countries. Increased food availability for us also increases stature and overall body size with marked increases in body weight from one generation to the next. Girls also become sexually mature at substantially earlier ages, which is at least partially caused by improved nutrition but likely has numerous other influences, including modern artificial lights and exposure to hormones and hormone mimics in food, plastics, and so forth, in addition to social stimulation of various sorts from dress fashion and peer influences to mass media and social networks. Despite precocious sexual activity, humans in developed countries have markedly lower fecundity compare to agricultural societies. Whether these similarities are systematic or accidental remains to be determined. The shared pattern is certainly suggestive of an adaptive response to increased food supply, increased survival rates, and higher population density.

In addition to these elaborate modifications in morphology and reproductive biology, the gregaria undergo profound changes in their social behavior that also offer a fascinating insight into the increased social complexity of urban life in general and that of developed countries in particular. Unlike fish schools, locust swarms are organized in the sense that they are based on long-standing mutual attraction that is analogous to clan affiliations of village-level human societies. Swarms also develop systematically or progressively.

Locust communities begin with bands of several thousand individuals whose behavior is organized or cohesive. Bands move together on the ground and generally prefer to move downhill, making their way around obstructions and covering as much as several miles per day. They pause to feed periodically before resuming their advance.

As they move, they encounter other bands and merge with them. They continue to act as cohesive units after they are fully mature and can fly. If an individual gets separated from the others while feeding, it rejoins the swarm whenever it moves overhead. The locusts settle in a curious rolling front. When the first insects land, others move on and settle in front of them. In this way, there are always different individuals at the leading edge. Much of their time is spent on the ground feeding and resting. When they move on, they often travel long distances, selecting areas where recent rain has produced plenty of lush green vegetation.

Although locusts were studied mainly because they were a costly threat that put many farmers out of business, they have much to tell us about how individuals are changed by transitioning from a solitary existence to large bands and enormous, well-organized swarms. They have much to tell us about ourselves as a species that began in very small migrating forager bands, before forming permanent settlements in villages of kin-affiliated individuals, before establishing extensive trading networks that stimulate settlement in larger towns and formation of central cities that contain most of the residents in developed countries. While the underlying mechanisms may be different, there are two striking similarities. One is the relative speed at which large social groups emerge. The other is their inherent instability, which is in marked contrast to the social insects that preserve their highly organized social life over millions of years.

Just as locust swarms are inherently unstable because they crash the local ecology, no urban civilization has stood the test of time according to George Tainter. Most are lucky to make it past their first thousand years, whether they succumb to ecological disaster, disease, or inherent problems of increasing economic inefficiency after the manner of ancient Rome. While they survive, they modify human behavior in curious ways that are startlingly similar to the transformation of locusts.

EFFECTS OF HYPERSOCIALITY ON HUMANS

All civilizations experience unusually rapid population growth at some point. They resemble locust swarms with their exponential growth of population, whether this is measured in terms of the overall size of the population or in terms of increasing population density.

Crowding alters the brain chemistry and general biology of grasshoppers. Modern humans are biologically very different from earlier populations also. We are taller, heavier, longer lived, and becoming increasingly intelligent thanks to the greater demands placed on our brains by contemporary hypersociability. These include literacy that remodels the functional anatomy of the brain to facilitate reading of text, and even changes the shape of eyeballs. The emergence of written language some five thousand years ago generated functional change in the form of a "letterbox" dedicated to word recognition. Adaptive brain changes in response to altered living conditions are nothing new, and they can implicate natural selection as well as developmental modification.

Not all changes in brain biology reflect social adaptation. Soon after the domestication of dogs, human sensory acuity declined, shrinking the thymus that processes sensory information, according to analysis of skull shape that indirectly reveals underlying brain structures, for example.[64] Of course, the larynx also changed in response to the evolution of spoken language, and functional areas of specialization emerged for speech production (Broca's area) and comprehension (Wernicke's area) although the localization of these functions excites ongoing controversy.

The watchword for modern human adaptation is change—change that neither ends nor slows down. Our behavioral plasticity renders bee societies an unworkable analogy given the stability of their social lives from millennium to millennium. Similarly, the genetic determinism of evolutionary psychology with its focus on static cross-cultural universals is out of touch with reality. Human behavior is not predetermined by genes any more than that of any other animal is: all behavioral phenotypes are the product of a complex developmental unfolding, and our psychology is constructed from our experiences rather than getting transmitted by some mythical genetic program that is overwhelmingly rejected by developmental geneticists.

Humans are perfected for handling change and make their way in a staggering variety of habitats, exploiting an endless array of foods and other resources. We are far more like locusts than we resemble bees. When the locust swarm overruns local habitats, it moves on to new places and new plant types, until everything gets desolated and the swarm moves on.

Human urban civilizations have a similar scorched-earth character. These proved unstable for a variety of reasons ranging from ecological unsustainability to political disintegration. Such problems are surfacing in the contemporary world with signs of ecological collapse related to overconsumption and climate change and accelerating inequality that promises to do for us what it did for the Romans. It would be comforting to imagine that complex human societies can be stable and enduring. History tells us otherwise. However large, powerful, complex, and confident a society grows, it comes with a limited shelf life. This was true of the Mesopotamians, the Egyptians, the Romans, the Ottomans, and the Incas, among many others. Can we beat the odds?

GLOBAL CAPITALISM

Human civilizations may be inherently unstable just as locust swarms are. Yet humans differ ecologically in being integrated across wide geographical regions. Viewed as a single entity, contemporary global capitalism is much larger than any preceding civilization, and it is far better integrated in terms of communications, infrastructure, and economic interdependence. Larger civilizations are more stable as a rule because they are insulated from regional ecological failures. Unfortunately, such regional risk sharing will be no help in the future because the entire globe is affected by ecological problems due to population growth and economic development. The dismal reality is that we are working our way through another boom cycle that will be followed by a spectacular bust. The only hope is that future generations confound informed prediction by resolving perennial problems of greed, climate change, and ecological degradation.

Like every other dominant civilization, we imagine that the world is improving with us and that our way of life will continue indefinitely into the future. Yet the one conclusion that we can draw from urban civilizations of the past is just the opposite. Like locusts, people have periodically assembled in large numbers from dispersed ancestral populations. Admittedly, urban civilizations lasted for many generations although all proved unstable to this point in history. This longevity is less impressive when considered as number of generations rather than years given that locusts breed on a time scale of a few months rather than decades.

Already, urban populations are taking over the entire globe with growing threats to global ecosystems. The global economy currently unites the activities of people all around the planet. We are close to the peak of population, just before the crash in numbers (foretold by extremely low fertility at half of replacement values in the most developed countries). [65]

Globalization is accompanied by a convergence of democratic political systems that unite developed countries, even though some large developing countries may appear headed in a different direction. These include the persistence of totalitarianism in China and Russia, but this is an anomaly and the eventual opening up of communications and the Internet will likely have the same consequences there as elsewhere around the globe. The tyranny of the regime will be exposed for all to see, complete with executive murders of dissidents, censorship of dissenting views, and gulags overflowing with innocent victims of state repression. Young defectors will leave in droves, preferring to live in countries that respect individual rights. Currently, dictatorships such as China and Russia use the Internet for controlling citizens and disseminating propaganda, however.

It is surprising, even shocking, that such totalitarian regimes have survived thus far, but there are no fully developed countries where such practices persist so that a wager on their indefinite continued existence has to be a losing bet. This is true even if economic problems in some developed countries periodically give rise to extreme right-wing political movements. The current wave of populist unrest in developed countries is connected with a decline in living standards of working people due to unemployment, lack of education and training, and declining pay in a global economy. Such phenomena are common during times of economic stress, as illustrated by fascism during the Great Depression, but they are mostly temporary and dissipate when prosperity returns. The broad arc of political change is a strong trend of increase in the number of countries that have adopted democratic government.

Leaving aside these political anomalies, the modern world is characterized by a remarkable, and unprecedented, evolutionary convergence. We see this in our physical and physiological adaptation. People in developed countries are taller, heavier, more intelligent, and more long-lived.

These biological adaptations to current conditions are mostly independent of gene selection, analogous to the adaptation of desert grasshoppers to rain in a formerly arid ecology. The human equivalent of rain is the continuous economic growth following the Industrial Revolution that liberated us from the Malthusian desert of all previous human societies. Ancient history and modern environmentalism each give good reasons for skepticism about the permanence of this shift, however.

We are altered not just in our somatic adaptation to better nutrition and improved health but also in our social and intellectual responses to the greater complexity, hypersociability, and extensive opportunities of modern life. Affluence increases intrinsic work motivation and the need for achievement so that there is a progressive element of always wanting to improve upon what went before. Malthusian societies were characterized by passivity, low ambition, stasis, and class inertia.

In a progressive environment, people focus on improving living conditions so as to satisfy higher aspirations as Maslow envisaged them in his hierarchy of needs. On the other hand, in Malthusian (or steady state) societies, there is much more of a focus on satisfying basic material needs low down in Maslow's hierarchy. The fact that such needs were hard to satisfy and that life was always precarious meant that human experience was invariably difficult and unpredictable. Religion flourished as a mechanism for dealing with emotional problems from fear, insecurity, bereavement, and other forms of misery, and there was far less interest in self-actualization and artistic expression.

Modern life is no panacea either. We experience many problems that were not salient a century ago. The Cold War created fear and uncertainty unlike anything seen before. Other uniquely modern problems range from climate change and global terrorism to extremely low fertility in developed countries (partly attributable to inflated costs of raising children in a modern service economy). Of course, there is minimal job security in some developed countries, including the United States. Then there are problems of worsening income inequality in some of the wealthiest countries, threatening political stability and national unity.

THE WORLD COMES TOGETHER

Whatever the problems of development, change in developing nations follows a predictable recipe. Developed countries converge on a similar societal prototype. This convergence is highlighted by increased ease of travel among affluent countries. We are unified by rapidly accelerating international trade; shared fads, foods, and fashions; and the fact that social media, entertainment, and sports transcend national boundaries. The conspicuous exceptions here are communist regimes like Russia and China that put restrictions on Internet use by citizens (something that affects as many as a third of Internet users worldwide).[66] Neither of these totalitarian regimes is yet fully developed, of course, and their repression and corruption could compromise their future development.

Even for a state as powerful as China, where censorship is referred to as "Internet sovereignty," it is difficult to regulate social media because the volume of information is so great. It is one thing to use a filter that blocks all mention of Tiananmen Square in search results (via the "Great Firewall") on Chinese Internet sites.[67] It is quite another to prevent citizens from accessing social media accounts of these events posted in other countries on popular sites such as YouTube, Twitter, or Instagram using virtual private networks that shield user identity. Indeed, some cybersecurity experts believe that the chilling effect of a belief that Big Brother is watching on the Internet is more useful to the Chinese regime than the filters themselves (in which American companies such as Amazon and Microsoft collaborate as a condition of doing business in China, Google having withdrawn from the country).

Tyrannies can persist for decades, yet the modern world mostly favors the free flow of information around the globe that has the same effect on totalitarianism as sunshine has on snow. So much energy is required to oppress citizens that totalitarianism is irrational, which may be why it persists only in communist regimes that have endless cheap manpower to throw at the problem.

While the Internet is also used by Russia and other countries to project power, spin propaganda, and conduct cyberwar, these activities continue in a world that is inadequately prepared for such intrusions of mendacity, just as it is unprepared for cyberfraud and theft. Yet we are alerted to these problems and may begin taking corrective measures.

Tiny Estonia was brought to its knees by a Russian cyberattack, in 2013, that compromised government Internet sites.[68] Estonians subsequently protected themselves by providing citizens with a smart e-card for logging onto the Internet and recruited a small army of cybercrime experts. Backing up government files was another key measure. The Estonians installed secure mirror sites using Blockchain technology so that their services, if compromised again, could be restored in moments. As a result of such efforts, their Internet is now the most secure in the world. Any other country could do the same if they made the necessary investment in electronic security.

Despite anomalies like Russia, societies around the world are converging in dramatic ways. This change is adaptive. "Adaptive" means that people are responding to environmental change in ways that improve their chances of leading long, healthy lives. The primary driver of change is not gene selection but competition between individuals. Competition is not restricted to mere survival but still involves men competing over women. Both sexes compete over wealth and social status. Why does this matter?

To begin with, these conclusions contradict virtually everything one might read about the causes of change in human societies. The two dominant schools of thought—cultural determinism and evolutionary psychology—reject such adaptive explanations of modern life. Neither of these approaches can accept that contemporary social change is adaptive, or a predictable response to contemporary environmental conditions. According to evolutionary psychologists, we are adapted to ancestral environments not contemporary ones so that gene-based adaptation to modern environments is played down. According to cultural determinists, social change is not produced by adaptation to changing environmental conditions but by "culture," an approach that is disturbingly lacking in scientific rigor as each society dances to its own drummer, so to speak, with idiosyncratic variation predominating and unpredictable change over time further muddying the water. Both of these approaches are ill-equipped to grasp the many ways in which humans respond to the environment of continuous economic growth that most advanced countries have experienced over the past century.

Convergent modern societies are every bit as much the product of prosperity as locust swarms are the product of rain. We change in response to better nutrition just as grasshoppers do, and we are affected by increased social stimulation in much the same ways as they are, namely through profound alterations of brain biology and social behavior. On the one hand, it might seem absurd to claim that an urban landscape full of skyscrapers and cars bears any similarity with the activities of an invertebrate agricultural pest. Yet the level of technological complexity may not be all that important. What we are observing is a social species so biologically successful that it completely overwhelms its physical environment. The artificial desert of an asphalt parking lot and a field of corn that has been denuded of all green leaves are ecologically equivalent effects. We are seeing how coordinated masses of a species can stretch their ecosystem past its ability to recover.

Even technology is a numbers game because human technological development only happens when people get together in large populations, as illustrated by the rapid loss of boat-building technology by Tasmanian Islanders once they lost their land bridge (or isthmus) to the mainland, which shrank the population and deprived them of boat-building experts. This historical episode highlights the instability of complex societies but also suggests that the currently developing global economy will be much more stable than any previous complex society—at least to the extent that size ensures stability. Of course, the scale of our global economy also threatens global ecosystems with grave and pressing issues for the ecological viability of our high-consumption way of life.

Contrary to evolutionary psychologists and cultural determinists alike, a lot of contemporary social change is adaptively predictable and mediated by the same nongenetic mechanisms that shape the behavior of other species to match current conditions. Our lives since the Industrial Revolution involved a marked secular improvement in the quality of life of the overall population. This short period is exceptional and a break with everything that preceded it. In this environment, once our basic needs are met, we gravitate to derived needs based on intrinsic motivation and conspicuous consumption.

Although living conditions improved recently, it is wrong to interpret earlier history as a continuing trend of improvement. Early farmers were substantially more miserable than their hunter-gatherer forebears despite producing a lot more food and having appreciably more children with a substantial boom in population (dwarfed only by the current period of industrial farming and the associated population explosion). Agriculture must have been experienced as a step backward from a highly varied diet to a dangerously restricted one and from a leisured lifestyle to a regimen of heavy labor that produced repetitive-stress-related injuries. [69]

In health terms, the first real improvement in human affairs came late in the nineteenth century in England, and improvements in general health of the population came several decades after factory production was well established and large proportions of the current population inhabited industrial centers. The unevenness of early improvements in health are explainable in terms of challenges posed by the urban environment, such as polluted, unsanitary housing conditions where crowding contributed to epidemics of cholera and other infectious diseases. Increasing wages did not immediately boost health, and the latent promise of the Industrial Revolution emerged only after the development of public sewer systems, the provision of clean public drinking water, and the emergence of more sterile treatment conditions in hospitals that greatly reduced the number of women dying in childbirth.

Real progress in health in the second half of the nineteenth century is illustrated by increasing stature of military recruits. [70] There were also substantial improvements in life expectancy. [71] Of course, these trends would really come into their own during the twentieth century: improvements in both measures of general health were unlike anything ever seen before on the planet in any society, much less as a global trend.

For several thousand years, our forager ancestors evaded the sad horror of work on a farm. They knew what it took to cultivate plants but had no desire to get into this arduous occupation as a full-time job until forced into it by the Pleistocene Overkill. This contradicts the popular narrative of agriculture as an improvement in the quality of life facilitated by discovery of how to cultivate crops.

Whether agriculture was good or bad for the quality of life, it was a huge success in boosting human populations, giving us the status of most successful primate on the planet in Darwinian terms. This facilitated the emergence of urban civilizations, ancient as well as modern. Green agriculture exaggerated population increases by reducing hunger in poor countries like India. Improvements in medical science also boosted life expectancy. So the current phase of human evolution—the population explosion—created an unprecedented swarming of our species into the world's cities.

THE HUMAN SWARM

The brain function and behavior of a simple animal like the desert locust is affected profoundly by what other locusts are doing. It is hardly surprising that the same would be true of highly social animals like most mammals. Enrichment experiments show that rats learn complex mazes more rapidly if they are housed socially rather than if they are caged alone.[72] Similarly, human intelligence is affected by what other people are doing.

In modern environments, people are exposed to more contact with others whether by direct interaction or through communication media and entertainment. Consequently, we are better at processing complex information quickly, that is, develop higher intelligence, a phenomenon known as the Flynn effect.

With universal education and more leisure reading, the volume of information accessed increased dramatically. Urban life made increasing cognitive demands from mere navigation in a complex maze of streets to acquiring complex work procedures and adjusting to a more diverse social landscape wherein individuals think and act differently.

In the digital world, we experience ever more contact with other people and process increasing amounts of information at ever faster speeds, whether for social or occupational purposes. The digital world offers an unlimited diet of information and entertainment that challenges our brains during most of our waking hours in contrast with the far more monotonous existence of preceding generations.[73] IQ scores rise. Scholars concluded that this was not primarily due to the storing of large quantities of information but rather to an enhancement of our generalized capacity to

solve problems.[74] In other words, we are better adapted to an information-rich contemporary environment.

Economic development works rather like a brain enrichment program:
it affects our brains in profound ways. Similarly, increasing population
density modifies the nervous systems of short-horned grasshoppers.
Grasshopper adaptation to an abundance of food affects feeding and sexual behavior in remarkable ways. The human abundance stemming from
economic development also affects both our feeding system and our sexual behavior, as illustrated by the rise of premarital sexuality and obesity.
Although we may be inclined to interpret these issues in terms of high-
level processes of social influence, it is refreshing to think about them as
low-level adaptation to environmental change analogous to the grasshopper to locust transition.

Swarming is what makes a grasshopper a locust. Interestingly, swarming is determined by changes in individuals. The same principle holds
widely in nature. Whether it is the efficient movement of ants on a trail
(that beats the performance of human vehicular traffic) or the beautiful
coordinated movements of fish schools or flocks of starlings "dancing" in
the evening skies, these phenomena appear holistic or centralized. Yet
such feats of coordinated movement have neither a conductor nor a score,
so to speak. Instead, they are an emergent property of low-level adjustment by constituent individuals.[75]

Among schooling fish, for instance, individual fish modify their
movements in accordance with pressure sensations received at their lateral line organs and move in the right direction to remain close to the other
members of the school without colliding with them. No doubt, individual
gene selection played a role in maintaining such movement patterns including their sensory and motor basis. Analogous adjustments keep birds
in large flocks from colliding in midair so that they trace out patterns in
the air that are as coordinated as fish schools.

It is tempting to imagine that a flock of starlings in patterned flight in
the evening sky is answering to some centralized authority, some alpha
bird who calls the dance, so to speak. Yet there is no known centralized
coordination. Like schooling fish, the pattern of coordinated flight is
entirely bottom-up. It is determined by the responses of individuals to the
movements of their immediate neighbors. The pattern we see in coordinated flying is therefore an emergent property of the low-level responses
of individuals. Similarly, the tendency of migrating geese to fly in V
formations evidently reflects the aerodynamic advantage to following
individuals riding in the wake of a front bird.

It is easy to stretch such analogies too far for human social behavior,
however. We *do* respond to centralized authority of governments, and we
are affected by communication patterns that have centralized features,

whether it is the stringing of phone lines from cities to towns or the dissemination of information through newspapers and TV that is both centralized and biased in favor of owners and political elites. Even famously democratic social media give a lot more influence to celebrities than others. Despite some centralization of power and influence in government and mass media, many features of human social behavior remain distinctly bottom up.

THE NUTS AND BOLTS OF SOCIAL INFLUENCE

Like schooling fish, our actions are strongly affected by what people around us are doing. This is nicely illustrated by the sexual behavior of unmarried women. In some societies, the majority of women are sexually active before marriage. In others, sexual intercourse before marriage is treated as a capital offense. In such "societies of honor," women are penalized heavily if they are suspected of yielding to sexual temptation.[76] Being sexually active also damages their reputation and impairs their marriage prospects.

In more liberal societies, the majority of women are open to sexual behavior before marriage. There is no penalty for sexual expression in a committed relationship. Indeed, women who remain virgins are at a romantic disadvantage and get passed over in favor of more hedonistic competitors.[77]

Social scientists are fond of discussing such societal differences in terms of values, sexual morality, laws, and cultural traditions. They are more elegantly explained in terms of competition between individual women and their rivals over the affections of desirable men. Referring to values adds nothing that one could not already glean from analyzing the competitive behavior of individuals. They need no top-down regulation any more than a school of fish needs a choreographer. Residents of literate societies *are* constrained by legal traditions, of course. This is certainly true of fundamentalist Islamic societies where Sharia law sanctions death by stoning as the appropriate penalty for extramarital sex.

Even so, laws and traditions do not prevent sexual behavior from adapting to changes in the ratio of males to females.[78] In many African countries, repressive traditions, such as infibulation and clitoridectomy, are designed to curtail female sexual expression and prevailing religions forbid extramarital sexuality. Yet extramarital sex is common as illustrated by high HIV/AIDS infection rates among heterosexuals. For similarly impoverished countries in the Middle East, HIV/AIDS infection rates are much lower. The most plausible explanation for greater sexual expression in Africa compared to the Middle East is the fact that there are fewer men as a proportion of women. Women compete with others by offering sexual intimacy early in a relationship as opposed to a coy strategy of withholding sex until marriage that works in the Middle East because women are in greater demand.

If sexual behavior is consistent with widely held religious ideas, cultural determinists have a field day: they claim that the religious belief is responsible for the observed behavior. When that does not work out, they are far less willing to acknowledge a failure of prediction. Belonging to a religion that endorses polygamy does not make polygamous marriage any more or less likely in practice according to my country comparison based on Demographic and Health Surveys conducted in less-developed countries.[79] If cultural determinism cannot be falsified by such evidence, then it is not a falsifiable scientific theory but a rigid belief system.

How does one make sense of the wide-ranging impact of sex ratios on sexual psychology and behavior? Such phenomena cannot be explained in terms of traditional values, religion, government, or any other top-down phenomenon. Instead, it is a matter of human sexual behavior adapting to what others in the population are doing. So female students on one campus behave very differently from those on another based upon the supply of men, and these differences have nothing to do with what students believe about sexual morality on the first day of college classes.[80]

When sexual behavior (or any other social phenomenon) is a product of the limited choices available to individuals, this is methodological individualism. Methodological individualism is being recognized in animal behavior as the basis for fish schooling, for example. It has received far too little attention for humans, diverging as it does from favored top-down theories like cultural determinism.

Social changes that seem coordinated, centralized, or top-down may be nothing of the sort. Attitudes do change in lockstep with changing social behavior but that does not mean that one causes the other any more than the noise of a lawnmower is what makes it move. Instead, much social change is driven by adaptation of individuals to altered social and ecological realities. One key change was the Industrial Revolution when people suddenly became much more productive than ever before—a transition that echoes the Agricultural Revolution when food production increased markedly kicking off a population explosion.

Contemporary views of the Industrial Revolution seem flawed. We often hear that factories dehumanized people and forced them to work insanely hard. This view is questionable. The invention of steam looms did not force weavers to work harder and become more productive. If anything, hard work and productivity were, and are, internal, or psychologically driven. Employees worked harder because they were intrinsically motivated due to prospects of improved living conditions and social mobility. That much social change is both individually adaptive and triggered by human competition is a new way of thinking about history. Yet it helps to make sense of a lot of evidence that cannot be understood in other ways.

NOTES

I. THE VITAL ALTERNATIVES

1. Henry H. Goddard, *The Kallikak Family: A Study in the Heredity of Feeble Mindedness* (New York: MacMillan, 1912).

2. S. Scarr, R. A. Weinberg, and I. D. Waldman, "IQ Correlations in Transracial Adoptive Families," *Intelligence* 17 (1993): 541–55.

3. Frans Boas, "Language and Thought," in *Handbook of American Indian Languages* (Lincoln, NE: Bison Books, 1911); C. N. Degler, *In Search of Human Nature: The Decline and Revival of Darwinism in American Social Thought* (New York: Oxford University Press, 1991).

4. Joseph Lopreato and Timothy Crippen, *Crisis in Sociology: The Need for Darwin* (New Brunswick, NJ: Transaction, 1999); Steven Pinker, *The Blank Slate* (New York: Viking, 2002).

5. Nigel Barber, *The Science of Romance* (Buffalo, NY: Prometheus, 2002).

6. Derek Freeman, *The Fateful Hoaxing of Margaret Mead: A Historical Analysis of her Samoan Research* (Boulder, CO: Westview, 1999).

7. Donald Symons, *The Evolution of Human Sexuality* (New York: Oxford University Press, 1979).

8. J. Berger, J. E. Swenson, and I. L. Persson, "Recolonizing Carnivores and Naïve Prey: Conservation Lessons from Pleistocene Extinctions," *Science* 291 (2001): 1036–39.

9. Gilbert Gottlieb, *Individual Development and Evolution* (Mahwah, NJ: Lawrence Erlbaum, 2002).

10. Michael S. Blumberg, *Basic Instinct of Behavior* (New York: Thunder's Mouth Press, 2005), 15–16.

2. MECHANISMS OF CHANGE

1. Charles Darwin, *On the Origin of Species* (New York: Penguin, 1859/ 1982).

2. Michael Ruse, *Darwinism Defended: A Guide to the Evolution Controversies* (London: J. Johnson, 1798).

3. Thomas R. Malthus, *An Essay on the Principle of Population* (London: J. Johnson, 1798).

4. Ruse, *Darwinism Defended*, 64.

5. Blumberg, *Basic Instinct*.

6. David S. Moore, *The Developing Genome: An Introduction to Behavioral Epigenetics* (New York: Oxford University Press, 2015).

7. D. W. Pfennig, "Kinship and Cannibalism," *Bioscience* 47 (1997): 667–75.

8. T. B. Franklin et al., "Epigenetic Transmission of the Impact of Early Stress across Generations," *Biological Psychiatry* 68 (2010): 408–15.

9. J. Weiner, *The Beak of the Finch: A Story of Evolution in Our Time* (New York: Vintage, 1995).

10. S. Wilde et al., "Direct Evidence of Positive Selection of Skin, Hair, and Eye Pigmentation during the Last 5,000 y," *Proceedings of the National Academy of Sciences* 111 (2014): 4832–37.

11. Joseph Henrich, *The Secret of Our Success: How Culture Is Driving Human Evolution, Domesticating Our Species, and Making Us Smarter* (Princeton, NJ: Princeton University Press, 2015).

12. Blumberg, *Basic Instinct*.

13. F. John Odling-Smee, K. N. Laland, and M. W. Feldman, *Niche Construction: The Neglected Process in Evolution* (Princeton, NJ: Princeton University Press, 2003).

14. Nigel Barber, "Educational and Ecological Correlates of IQ: A Cross-National Investigation," *Intelligence* 33 (2005): 273–84.

15. Frank W. Marlowe, "Hunting and Gathering: The Human Sexual Division of Foraging Labor," *Cross-Cultural Research* 41 (2007): 170–95.

16. J. Berger et al., "Recolonizing Carnivores and Naïve Prey."

17. Franklin et al., "Epigenetic Transmission."

18. Moore, *The Developing Genome*.

19. V. Delaney-Black et al., "Violence Exposure, Trauma, and IQ and/or Reading Deficits among Urban Children," *Archives of Pediatric and Adolescent Medicine* 156 (2002): 280–85.

20. Manuel Velasquez-Manoff, *An Epidemic of Absence* (New York: Scribner, 2012).

21. I. Morgan and K. Rose, "How Genetic Is School Myopia?" *Progress in Retinal and Eye Research* 24 (2005): 1–38.

22. Barber, "Educational and Ecological Correlates of IQ."

23. A. Case and C. Paxon, "Stature and Status: Height, Ability and Labour Market Outcomes," *Journal of Political Economy* 116 (2008): 491–532.

24. Moore, *The Developing Genome*; R. Wilkinson and K. Pickett, *The Spirit Level: Why Greater Equality Makes Societies Stronger* (New York: Bloombury Press, 2010).

25. E. Novarro et al., "Prenatal Nutrition and the Risk of Adult Obesity," *Journal of Nutrition and Biochemistry* 39 (2017): 1–14.

26. Leda Cosmides and John Tooby, "From Evolution to Behavior: Evolutionary Psychology as the Missing Link," in *The Latest on the Best: Essays on Evolution and Optimality*, ed. John Dupre (Cambridge, MA: MIT Press, 1987), 277–306.

27. E. A. Maguire et al., "Navigation-Related Structural Change in the Hippocampi of Taxi Drivers," *Proceedings of the National Academy of Sciences* 97, no. 8 (2000): 4398–403, https://doi.org/10:1073/pnas.070039597.

28. Scan B. Carroll, *Endless Forms Most Beautiful: The New Science of Evo Devo and the Making of the Animal Kingdom* (New York: W. W. Norton, 2005).

29. H. C. Kemelrijk and H. Hildenbrandt, "Schools of Fish and Flocks of Birds: Their Shape and Internal Structure by Self-Organization," *Interface Focus* 2 (2012): 726–37.

30. A. Karmiloff-Smith, "An Alternative to Domain-General or Domain-Specific Frameworks for Theorizing about Evolution and Ontogenesis," *AIMS Neuroscience* 2, no. 2 (2015): 91–104, https://doi.org/10.3934/Neuroscience. 2015.2.91.

31. Gottlieb, *Individual Development and Evolution*.

32. B. G. Galef Jr., "Diving for Food: Analysis of a Possible Case of Social Learning in Wild Rats (*Rattus norvegicus*)," *Journal of Comparative Psychology* 95 (1980): 615–22.

33. Peter J. Richerson and Robert Boyd, *Not by Genes Alone: How Culture Transformed Human Evolution* (Chicago: University of Chicago Press, 2004).

34. Pfennig, "Kinship and Cannibalism."

35. D. Nettle, "The Evolution of Personality Variation in Humans and Other Animals," *American Psychologist* 61 (2006): 622–31.

36. E. Avital and E. Jabonka, *Animal Traditions: Behavioural Inheritance in Evolution* (Cambridge, UK: Cambridge University Press, 2000).

37. Symons, *The Evolution of Human Sexuality*.

38. Richard Dawkins, *The Selfish Gene* (New York: Oxford University Press, 1976).

39. N. E. Colias and C. Colias, *Nest Building and Bird Behavior* (Princeton, NJ: Princeton University Press, 1984).

40. Nigel Barber, "Explaining Cross-National Differences in Fertility: A Comparative Approach to the Demographic Shift," *Cross-Cultural Research* 44 (2010): 3–22; Hillard Kaplan et al., "Evolutionary Approach to Below Replacement Fertility," *American Journal of Human Biology* 14 (2002): 233–56; Joel Kotkin, *The Rise of Post-Familialism*, (Singapore: Civil Service College, 2012), http://www.cscollege.gov.sg/Knowledge/Pages/The-Rise-of-Post-Familialism. aspx; Paul W. Turke, "Which Humans Behave Adaptively, and Why Does It Matter?" *Ethology and Sociobiology* 11 (1990): 305–39.

41. G. R. Bentley, T. Goldberg, and G. Jasienska, "The Fertility of Agricultural and Non-Agricultural Traditional Societies," *Population Studies* 47 (1993): 269–81.

42. K. R. Fontaine et al., "Years of Life Lost due to Obesity," *JAMA* 289 (2003): 187–93.

43. Henrich, *The Secret of Our Success*.

44. Blumberg, *Basic Instinct*; Michael S. Blumberg, *Freaks of Nature: What Anomalies Tell Us about Development* (New York: Oxford University Press, 2009).

45. Richerson and Boyd, *Not by Genes Alone*.

46. Ronald Inglehart and Christian Welzel, *Modernization, Cultural Change, and Democracy* (Cambridge, UK: Cambridge University Press 2005).

47. Charles Francis Darwin, *Descent of Man and Selection in Relation to Sex* (London: Murray, 1871).

3. EVOLUTION AS A
DEVELOPMENTAL PROCESS

1. Edward O. Wilson, *Sociobiology: The New Synthesis* (Cambridge, MA: Belknap, 1975).

2. Blumberg, *Basic Instinct*.

3. Cosmides and Tooby, "From Evolution to Behavior."

4. Henrich, *The Secret of Our Success*.

5. Carroll, *Endless Forms Most Beautiful*.

6. David M. Buss, *Evolutionary Psychology: The New Science of the Mind* (Boston, MA: Allyn and Bacon, 1999).

7. Adam Smith, *The Wealth of Nations: A Translation into Modern English* (Manchester, UK: Industrial Systems Research, 1776/2015).

8. S. O. Lilienfeld and H. Arkowitz, "Are Men the More Belligerent Sex?" *Scientific American* (May/June 2010): 64–65.

9. C. Holden and R. Mace, "Sexual Dimorphism in Stature and Women's Work," *American Journal of Physical Anthropology* 110 (1999): 27–45.

10. A. A. Macintosh, R. Pinhasi, and J. T. Stock, "Prehistoric Women's Manual Labor Exceeded That of Athletes, through the First 5500 Years of Farming in Central Europe," *Science Advances* 3, no. 11, eaao3893 (2017), https://doi.org/10.1126/sciadv.aao3893.

11. Becky Hart and Todd Risley, *Meaningful Differences in the Everyday Experience of Young American Children* (Baltimore, MD: Paul H. Brookes, 1995).

12. A. Tellegen et al., "Personality Similarity in Twins Reared Apart and Together," *Journal of Personality and Social Psychology* 54 (1998): 1031–39.

13. Robert Plomin, *Nature and Nurture: An Introduction to Human Behavioral Genetics* (Belmont, CA: Wadsworth, 1990).

14. Scarr et al., "IQ Correlations in Transracial Adoptive Families."

15. Nigel Barber, *Why Parents Matter: Parental Investment and Child Outcomes* (Westport, CT: Bergin and Garvey, 2000).

16. Moore, *The Developing Genome.*

17. Franklin et al., "Epigenetic Transmission of the Impact of Early Stress."

18. Delaney-Black et al., "Violence Exposure, Trauma, and IQ."

19. D. Maestripieri, "Early Experience Affects the Intergenerational Transmission of Infant Abuse in Rhesus Monkeys," *Proceedings of the National Academy of Sciences* 102 (2005): 9726–29.

20. Franklin et al., "Epigenetic Transmission of the Impact of Early Stress."

21. Ibid.

22. Ibid.

23. Moore, *The Developing Genome.*

24. Barber, *Why Parents Matter.*

25. Nigel Barber, "Evolutionary Explanations for Societal Differences and Historical Change in Violent Crime and Single Parenthood," *Cross-Cultural Research* 41 (2007): 123–48; Franklin et al., "Epigenetic Transmission of the Impact of Early Stress"; M. Kalinichev et al., "Long-Lasting Changes in Stress-Induced Corticosterone Response and Anxiety-Like Behaviors as a Consequence of Neonatal Maternal Separation in Long-Evans Rats," *Pharmacology, Biochemistry, and Behavior* 73 (2002): 131–40; Moore, *The Developing Genome*; M. H. Teicher et al. "Developmental Neurobiology of Childhood Stress and Trauma," *Psychiatric Clinics of North America* 25 (2002): 397–426.

26. Carl H. Nightingale, *On the Edge: A History of Poor Black Children and Their American Dreams* (New York: Basic, 1993).

27. N. Barber, "Countries with Fewer Males Have More Violent Crime: Marriage Markets and Mating Aggression," *Aggressive Behavior* 35 (2009): 49–56; Hart and Risley, *Meaningful Differences.*

28. Barber, *Why Parents Matter.*

29. Carroll, *Endless Forms Most Beautiful.*

30. C. P. Groves, "The Advantages and Disadvantages of Being Domesticated," *Perspectives in Human Biology* 4 (1999): 1–12.

31. F. W. Dove, "Artificial Production of the Fabulous Unicorn," *Scientific Monthly* 42, no. 5 (1936): 431–36.

32. Blumberg, *Basic Instinct.*

33. D. Viala, G. Viala, and N. Fayein, "Plasticity of Locomotor Organization of Infant Rabbits Spinalized Shortly after Birth," in *Development and Plasticity of the Mammalian Spinal Cord*, eds. M. E. Goldberger, A. Gorio, and M. Murray (Padova, Italy: Liviana Press, 1986), 301–10.

34. Blumberg, *Basic Instinct.*

35. Hans Kruuk, *The Spotted Hyena* (Chicago: University of Chicago Press, 1972).

36. Blumberg, *Basic Instinct.*

37. Ibid.

38. Barber, *The Science of Romance.*

39. A. Galani et al., "Androgen Insensitivity Syndrome: Clinical Features and Molecular Defects," *Hormones* 7 (2008): 217–29.

40. Ibid.

41. Ciani A. Camperio, "Genetic Factors Increase Fecundity in Female Maternal Relatives of Bisexual Men as in Homosexuals," *Sexual Medicine* 6 (2009): 449–55.

42. Hans J. Eysenck, *The Structure of Human Personality* (New York: Wiley, 1953).

43. A. W. Lukaszewski and C. von Rueden, "The Extroversion Continuum in Evolutionary Perspective," *Personality and Individual Differences* 77 (2015): 186–92.

44. M. Gurven et al., "The Evolutionary Fitness of Personality Traits in a Small-Scale Subsistence Society," *Evolution and Human Behavior* 35 (2014): 17–25; Lukaszewski and von Rueden, "The Extroversion Continuum."

45. Ibid.

46. Ibid.

47. Ibid.

4. HOW DIFFERENT ARE HUMANS, OBJECTIVELY SPEAKING?

1. J. A. M. Thompson, "Bonobos of the Luukuru Wildlife Research Project," in *Behavioral Diversity in Chimpanzees and Bonobos*, ed. C. Boesch, G. Hohmann, and L. Marchant (Cambridge, UK: University of Cambridge Press, 2002), 61–70.

2. Darwin, *Descent of Man*.

3. R. W. Wrangham, "Evolution of Coalitionary Killing," *Yearbook of Physical Anthropology* 42 (1999): 1–30.

4. Wolfgang Kohler, *The Mentality of Apes* (New York: Harcourt Brace, 1925).

5. R. Epstein et al., "'Insight' in the Pigeon: Antecedents and Determinants of an Intelligent Performance," *Nature* 308, no. 81 (1984): 61–62.

6. B. T. Gardner and R. A. Gardner, "Two-Way Communication with an Infant Chimpanzee," in *Behavior of Nonhuman Primates*, vol. 4., ed. A. M. Schrier and F. Stollnitz (New York: Academic, 1972), 117–84.

7. H. S. Terrace et al., "Can an Ape Create a Sentence?" *Science* 206 (1979): 891–902.

8. E. S. Savage-Rumbaugh and R. Lewin, *Kanzi: The Ape at the Brink of the Hyman Mind* (New York: Wiley, 1996).

9. John Alcock, *Animal Behavior*, 4th ed. (Sunderland, MA: Sinauer, 1989).

10. P. Marler, and H. Slabbekoorn, eds., *Nature's Music: The Science of Birdsong* (New York: Elsevier Academic Press, 2004).

11. Ibid.

12. Ibid.

13. Ibid.

14. I. Pepperberg, "Grey Parrots: Learning and Using Speech," in *Nature's Music: The Science of Birdsong*, ed. P. Marler and H. Slabbekoorn (New York: Elsevier Academic Press, 2004), 389–450.

15. P. Marler and P. Mundinger, "Vocal Communication in Birds," in *The Ontogeny of Vertebrate Behavior*, ed. H. Moltz, (New York: Academic Press, 1971), 389–450.

16. P. Laiola and J. L. Tella, "Erosion of Animal Cultures in Fragmented Landscapes," *Frontiers in Ecology and Environment* 5 (2007): 68–72.

17. R. B. Edgerton, *Sick Societies* (New York: Free Press, 1992).

18. J. Henrich, "Demography and Cultural Evolution, Why Adaptive Cultural Processes Produced Maladaptive Losses in Tasmania," *American Antiquity* 69 (2004): 197–214.

19. Marler and Mundinger, "Vocal Communication in Birds."

20. G. E. Gallup, "Self-Awareness in the Chimpanzee," *Science* 167 (1970): 86–87.

21. J. M. Plotnik, F. M. D de Waal, and D. Reise, "Self-Recognition in an Asian Elephant," *Proceedings of the National Academy of Sciences* 103 (2006): 45–52; H. Prior, A. Schwarz, and O. Gunturkun, "Mirror-Induced Behavior in the Magpie (*Pica pica*)," *PLOS Biology* 6, no. 8 (2008): e202, https://doi.org/10. 1371/journals.pbio0060202.

22. M. Bekoff, "Observations of Scent-Marking and Discriminating Self from Others by a Domestic Dog (*Canis familiaris*): Tales of Displaced Yellow Snow," *Behavioural Processes* 55, no. 2 (2001): 75–79, https://doi.org/10.1016/S0376-6357(01)00142-5.

23. Frans de Waal, *Are We Smart Enough to Know How Smart Animals Are* (New York: W. W. Norton, 2016).

24. Ibid.

25. S. F. Brosnan, and F. B. de Waal, "Monkeys Reject Unequal Pay," *Nature* 425, no. 6955 (2003): 297–99.

26. Desmond Morris, *The Biology of Art* (New York: Knopf, 1962).

27. Ibid.

28. Ibid.

29. D. Morris, "Can Jumbo Elephants Really Paint?" *Daily Mail*, February 21, 2009, http://www.dailymail.co.uk/sciencetech/article-1151283/Can-jumbo-elephants-really-paint-Intrigued-stories-naturalist-Desmond-Morris-set-truth. html.

30. S. Inoue and T. Matsuzawa, "Working Memory of Numerals in Chimpanzees," *Current Biology* 17 (2007): R1004–R1005.

31. R. M. Lanner, *Made for Each Other: A Symbiosis of Birds and Pines* (New York: Oxford University Press, 1996).

32. E. Jablonka and M. Lamb, *Evolution in Four Dimensions: Genetic, Epigenetics, Behavioral and Symbolic Variation* (Cambridge, MA: MIT/A Bradford Book, 2006), 201.

33. Project Nim, "Project Nim: A Chimp's Very Human Very Sad Life," July 20, 2011, http://www.npr.org/2011/07/20/138467156/project-nim-a-chimps-very-human-very-sad-life.

34. Henrich, *The Secret of Our Success*.

35. Alex Mesoudi, *Cultural Evolution: How Darwinian Theory Can Explain Human Culture and Synthesize the Social Sciences* (Chicago: University of Chicago Press, 2011).

36. M. Harris, *Cows, Pigs, Wars, and Witches: The Riddles of Culture* (New York: Penguin Random House, 1974).

37. Henrich, *The Secret of Our Success*.

38. "Greenland's Viking Settlers Gorged on Seals," *Science News*, November 19, 2012.

39. G. Stockinger, "Archeologists Find Clues to Viking Mystery," *Der Spiegel*, January 10, 2013, http://www.spiegel.de/international/zeitgeist/archeologist-uncover-clues-to-why-vikings-abandoned-greenland-a-876626.html.

40. J. Arneborg, N. Lynnerup, and J. Heinemeier, "Human Diet and Subsistence Patterns in Norse Greenland AD c. 980–AD c. 1450: Archeological Interpretations," *Journal of the North Atlantic* (2012): 119–33, http://www.doi.org/dx.doi.org/10.3721/037.004.s309.

41. Ibid.

42. Ibid.

43. Ibid.

44. Ibid.

45. Ibid.

46. J. Henrich et al., "'Economic Man' in Cross-Cultural Perspective: Behavioral Experiments in 15 Small-Scale Societies," *Behavioral and Brain Sciences* 28 (2005): 795–855.

47. Mesoudi, *Cultural Evolution*.

48. Moore, *The Developing Genome*.

49. Mesoudi, *Cultural Evolution*.

50. M. W. Morris and K. Peng, "Culture and Cause: American and Chinese Attributions for Social and Physical Events," *Journal of Personality and Social Psychology* 67 (1994): 949–71.

51. J. Allik and R. R. McCrae, "Toward a Geography of Personality Traits," *Journal of Cross-Cultural Psychology* 35 (2004): 13–28.

52. G. H. Hodgson and T. Knudsen, *Darwin's Conjecture: The Search for General Principles of Social and Economic Evolution* (Chicago: University of Chicago Press, 2010).

53. Mesoudi, *Cultural Evolution*.

5. THE EMERGENCE OF
COMPLEX SOCIETIES

1. Y. Melamed et al., "The Plant Component of an Achulean Diet at Geshe-Benot Ya'aqov, Israel," *Proceedings of the National Academy of Sciences* 113, no. 51 (2016), https://www.doi.org/10.1073/pnas.160787211.

2. Nigel Barber, *Why Atheism Will Replace Religion: The Triumph of Earthly Pleasures over Pie in the Sky* (self-published, Amazon Digital Services, 2012),

Kindle, http://www.amazon.com/Atheism-Will-Replace-Religion-ebook/dp/B00886ZSJ6/.

3. Henrich, *The Secret of Our Success*.

4. R. G. Bednarik, "An Experiment in Pleistocene Seafaring," *The International Journal of Nautical Archeology* 27 (1998): 139–49.

5. "Neanderthals Mated with Modern Humans Much Earlier Than Previously Thought, Study Shows," *Science Daily*, February 17, 2016, https://www.sciencedaily.com/releases/2016/02/160217140302.htm?utm_source=feedburner&utm_medium+feed&utm_campaign=Feed%253A+sciencedaily%252Ffossils_ruins%252Farchaeology+%2528Archaeology+News+--+ScienceDaily%2529.

6. P. Khaitovich et al., "Metabolic Changes in Schizophrenia and Human Brain Evolution," *Genome Biology* 9, no. R124 (2008): 1–11; R. Nixon, "Cooking and Cognition: How Humans Got So Smart," *LiveScience* August 11, 2008, 11 August.

7. A. L. File, G. P. Murphy, and S. A. Dudley, "Fitness Consequences of Plants Growing with Siblings," *Proceedings in Biological Science* 279, no. 1727 (November 3, 2011): 209–18, https://www.doi.org.10,1098/rsph.2011.1995.epub2011nov9.

8. E. Callaway, "Blocking 'Happiness' Chemical May Prevent Locust Plagues," *New Scientist*, January 29, 2009, http://www.newscientist.com/article/dn16505-blocking-happiness-chemical-may-prevent-locust-plagues.html#,VN4Z_fnF9sE.

9. Steven Johnson, *Everything Bad Is Good for You: How Today's Popular Culture Is Actually Making Us Smarter* (New York: Riverhead, 2006).

10. W. T. Dickens and J. R. Flynn, "Heritability Estimates versus Large Environmental Effects: The IQ Paradox Resolved," *Psychological Review* 108 (2001): 346–69.

11. Hemelrijk and Hildenbrandt, "Schools of Fish and Flocks of Birds."

12. M. Shostak, *Nisa: The Life and Words of a !Kung Woman* (Cambridge, MA: Harvard University Press, 1981).

13. A. W. Johnson, and T. Earle, *The Evolution of Human Societies*, 2nd ed. (Stanford, CA: Stanford University Press, 2000).

14. . M. Sahlins, "Notes on the Original Affluent Society," in *Man the Hunter*, ed. Richard B. Lee and Irven DeVore (New York: Aldine, 1968), 85–89.

15. C. R. Ember and M. Ember, "Resource Unpredictability, Mistrust and War: A Cross-Cultural Study," *The Journal of Conflict Resolution* 36 (1992): 242–62.

16. Symons, *The Evolution of Human Sexuality*.

17. Colin Rudge, *Neanderthals, Bandits and Farmers: How Agriculture Really Began* (New Haven, CT: Yale University Press, 1999).

18. Wilkinson and Pickett, *The Spirit Level*.

19. Christopher Boehm, *Hierarchy in the Forest* (Cambridge, MA: Harvard University Press, 2000).

20. R. A. Bentley et al., "Community Differentiation and Kinship among Europe's First Farmers," *Proceedings of the National Academy of Sciences* 109, no. 24 (2012): 9326–30, https://www.doi.org/10.1073/pnas.1113710100.

21. Napoleon A. Chagnon, *Yanomamo: The Fierce People* (New York: Holt, Rinehart, & Winston, 1968).

22. J. E. McClellan and H. Dorn, *Science and Technology in World History: A Thousand-Year History* (Cambridge, MA: MIT Press, 1999).

23. E. Slingerland, J. Henrich, and A. Norenzayan, "The Evolution of Prosocial Religions," in *Cultural Evolution: Society, Technology, Language, and Religion*, ed. Peter J. Richerson and Christiansen H. Morten (Cambridge, MA: The MIT Press, 2013), 335–48.

24. R. Sosis and E. R. Bressler, "Cooperation and Commune Longevity: A Test of the Costly Signaling Theory of Religion," *Cross Cultural Research* 37 (2003): 211–39.

25. R. Sosis and B. J. Ruffle, "Religious Ritual and Cooperation: Testing for a Relationship on Israeli Religious and Secular Kibbutzim," *Current Anthropology* 44 (2003): 713–22.

26. P. Osborne, "Saudi Arabia Brutality against People Revealed in Video," *Daily Mail Online*, April 5, 2007, http://www.dailymail.co.uk/news/article-3502079/Saudi-Arabia-s-kingdom-savagery-DOES-Britain-cosy-butchers.html.

27. Robert Cialdini, *Influence: Science and Practice*, 2nd ed. (Glenview, IL: Scott Foresman, 1988).

28. R. Floud et al., *The Changing Body: Health, Nutrition, and Human Development in the Western World Since 1700* (Cambridge, UK: NBER/Cambridge University Press, 2011).

29. E. Benson, "Marsupial Lion's Primate-Like Forearms Made It a Unique Predator," *New Scientist*, August 17, 2016, https://www.newscientist.com/article/2101507-marsupial-lions-primate-like-forearms-made-it-a-unique-predator/.

30. Matt Ridley, *The Rational Optimist* (New York: Harper Collins, 2010), 20–21.

31. Floud et al. *The Changing Body*.

32. Richerson and Boyd, *Not by Genes Alone*.

33. H. Dreher, "Why Did the People of Roseto Live So Long?" *Natural Health* 23, no. 5 (1993): 72–83.

34. N. Barber, "Why Behavior Matches Ecology: Adaptive Variation as a Novel Approach," *Cross-Cultural Research* 49 (2015): 57–89.

35. Barber, *Why Parents Matter*.

36. Pinker, *The Blank Slate*.

37. Inglehart and Welzel, *Modernization, Cultural Change*.

38. E. Cashdan, "Attracting Mates: Effects of Paternal Investment on Mate Attraction Strategies," *Ethology and Sociobiology* 14 (1993): 1–24.

39. C. Goldin, "Career and Family: College Women Look to the Past," Working Paper #5188 (Cambridge, MA: National Bureau of Economic Research, 1995).

40. Eric Klinenberg, *Going Solo: The Extraordinary Rise and Surprising Appeal of Living Alone* (New York: Penguin, 2012).

41. T. Caplow, L. Hicks, and B. Wattenberg, *The First Measured Century: An Illustrated Guide to Trends in America, 1900–2000* (La Vergne, TX: AEI Press, 2001).

6. A DARWINIAN TAKE ON THE INDUSTRIAL REVOLUTION

1. Wilson, *Sociobiology*.

2. Darwin, *Descent of Man*.

3. R. H. Frank, T. Gilovich, and D. T. Regan, "Docs Studying Economics Inhibit Cooperation?" *Journal of Economic Perspectives* 7 (1993): 159–71.

4. O. Galor and O. Moav, "Natural Selection and the Origin of Economic Growth," *Quarterly Journal of Economics* 117 (200): 1133–91.

5. Lopreato and Crippen, *Crisis in Sociology*.

6. Blumberg, *Basic Instinct*.

7. Henrich, *The Secret of Our Success*.

8. Bednarik, "An Experiment in Pleistocene Seafaring."

9. R. Cameron and L. Neal, *A Concise Economic History of the World*, 4th paperback ed. (New York: Oxford University Press, 2003), 21–23.

10. William J. Bernstein, *A Splendid Exchange: How Trade Shaped the World* (New York: Atlantic Monthly Press, 2008), 66.

11. Lincoln P. Paine, *The Sea and Civilization: A Maritime History of the World* (New York: Alfred A. Knopf, 2013).

12. Angus Maddison, *The World Economy: A Millennial Perspective* (Paris: OECD, 2001).

13. J. B. DeLong, *Estimates of World GDP, One Million B.C—Present* (Berkeley, CA: UC–Berkeley Department of Economics, 1998), http://www.delong.typepad.com/print/20061012_LRWGDP.pdf.

14. Bernstein, *A Splendid Exchange*.

15. Ibid.

16. Paine, *The Sea and Civilization*.

17. Stockinger, "Archeologists Find Clues to Viking Mystery."

18. Rose George, *Ninety Percent of Everything* (New York: Metropolitan Books, 2013).

19. Rudge, *Neanderthals, Bandits and Farmers*.

20. Floud et al., *The Changing Body*.

21. G. Hillman et al., "New Evidence of Late Glacial Cereal Cultivation of Abu Hureyra on the Euphrates," *The Holocene* 11, no. 4 (2001): 383–93.

22. Johnson and Earle, *The Evolution of Human Societies*.

23. D. Q. Fuller, G. Willcox, and R. G. Allaby, "Early Agricultural Pathways: Moving outside the 'Core Area' Hypothesis in Southwest Asia," *Journal of Experimental Botany* 63 (2012): 617–33.

24. I. De Garine, "The Trouble with Meat: An Ambiguous Food," *Estudio del Hombre* 19 (2005): 33–54.

25. Rudge, *Neanderthals, Bandits and Farmers*.

26. A. M. Rosen and I. Rivera-Collazo, "Climate Change, Adaptive Cycles, and the Persistence of Foraging Economies during the Late Pleistocene/Holocene Transition in the Levant," *Proceedings of the National Academy of Sciences* 109 (2012): 3640–45.

27. Floud et al., *The Changing Body*.

28. Ridley, *The Rational Optimist*, 35.

29. Maddison, *The World Economy*.

30. Mary S. Hartman, *The Household and the Making of History* (Cambridge, UK: Cambridge University Press, 2004).

31. Nigel Barber, *The Myth of Culture: Why We Need a Genuine Natural Science of Societies* (Newcastle-upon-Tyne, UK: Cambridge Scholars Press, 2008); N. Barber, "Explaining Cross-National Differences in Fertility: A Comparative Approach to the Demographic Shift," *Cross-Cultural Research* 44 (2010): 3–22; V. K. Oppenheimer, "Women's Rising Employment and the Future of the Family in Industrial Societies," *Population and Development Review* 20 (1994): 293–42.

32. David Landes, *The Wealth and Poverty of Nations* (London: Little Brown, 1999): .8.

33. Galor and Moav, "Natural Selection and the Origin of Economic Growth."

34. Robert Kurzwell, *The Singularity Is Near* (New York: Viking/Penguin, 2005).

35. Geoffrey Clark, *A Farewell to Alms: A Brief Economic History of the World* (Princeton, NJ: Princeton University Press, 2007).

36. Ibid.

37. Martin Ford, *Rise of the Robots: Technology and the Threat of a Jobless Future* (New York: Basic, 2015).

38. David C. McClelland and David C. Winter, *Motivating Economic Achievement* (Toronto, Ontario: Collier-Macmillan, 1969).

39. Beatrice B. Whiting and John B. Whiting, *Children of Six Cultures* (Cambridge, MA: Harvard University Press, 1975).

40. D. M. Wegner and D. Schaefer, "The Concentration of Responsibility," *Journal of Personality and Social Psychology* 36 (1978): 147–55.

41. Alexander V. Chayanov, *The Theory of Peasant Economy* (Madison: The University of Wisconsin Pres, 1986).

42. Case and Paxon, "Stature and Status."

43. G. Hitsch, A. Hortacsu, and D. Ariely, "What Makes You Click? Mate Preferences in Online Dating," *Quantitative Marketing and Economics* 8 (2010): 393–27.

44. Barber, *The Science of Romance*.

45. Marcia Guttentag and Paul Secord, *Too Many Women: The Sex Ratio Question* (Beverly Hills, CA: Sage. 1983).

46. R. Mace, F. Jordan, and C. Holden, "Testing Evolutionary Hypotheses about Human Biological Adaptation Using Cross-Cultural Comparison," *Com-*

parative Biochemistry and Physiology Part A: Molecular and Integrative Physiology 136, no. 1 (2003): 85–94.

47. Steven S. Pinker, *The Better Angels of Our Nature: Why Violence Has Declined* (New York: Viking Penguin, 2011).

48. N. Barber, "Creative Productivity and Marriage Markets: Mating Effort and Career Striving as Rival Hypotheses," *Journal of Genius and Eminence* 2 (2017): 32–44, https://www.doi.org/10.18536/jge.2017.04.02.01.04.

49. Q. Jiang and J. J. Sanchez-Barricarte, "Bride Price in China: The Obstacle to 'Bare Branches' Seeking Marriage," *The History of the Family* 17 (2012): 2–15.

50. B. Low, "Cross-Cultural Patterns in the Training of Children," *Journal of Comparative Psychology* 103 (1989): 311–19.

51. Klinenberg, *Going Solo*.

52. Landes, *The Wealth and Poverty of Nations*, 9.

53. N. Barber, "Explaining Country Differences in Productivity: A Work-Motivation Approach to the Industrial Revolution" (under review).

54. Caplow et al., *The First Measured Century*.

55. United Nations, *Sex Differences in Childhood Mortality* (New York: Department of Economic and Social Affairs, Population Division, 2011).

56. David Popenoe, *Disturbing the Nest: Family Change and Decline in Modern Societies* (Hawthorne, NY: Aldine de Gruyter, 1998).

57. United Nations, *Human Development Reports* (New York: United Nations, 2015), http://www.hdr.undp.org.

58. Robert J. Gordon, *The Rise and Fall of American Growth* (Princeton, NJ: Princeton University Press, 2016).

59. Ford, *Rise of the Robots*.

60. Barber, "Explaining Country Differences in Productivity."

61. S. Denyer, "China's Scary Lesson to the World: Censoring the Internet Works," *The Washington Post*, May 23, 2016, https://www.washingtonpost.com/world/asia_pacific/chinas-scary-lesson-to-the-world-censoring-the-internet-works/2016/05/23/413afe78-ff3-11e5-8bb1-f124a43f84dc_story.html.

62. T. Hesketh and Z. W. Xing, "The Effect of China's One-Child Family Policy after 25 Years," *New England Journal of Medicine* 353 (2005): 1171–76

63. "Data Mining Reveals the Extent of China's Ghost Cities," *MIT Technology Review*, November 2, 2015, https://www.technologyreview.com/s/543121/data-mining-reveals-the-extent-of-chinas-ghost-cities/.

64. Popenoe, *Disturbing the Nest*.

65. Craig C. Pinder, *Work Motivation: Theory, Issues, and Applications* (Glenview, IL: Scott, Foresman, 1984).

66. McClellan and Dorn, *Science and Technology in World History*.

67. Barber, "Explaining Country Differences in Productivity."

68. William J. Wilson, *When Work Disappears: The World of the New Urban Poor* (New York: Vintage, 1997).

69. Kotkin, *The Rise of Post-Familialism*.

70. Caplow et al., *The First Measured Century*.

71. K. Hennigan et al., "Impact of Introduction of Television on Crime in the United States," *Journal of Personality and Social Psychology* 42 (1982): 461–77.

72. L. Heller, "Most Desired Brands by Gender," *Daily Finance*, February 2, 2011, http://www.dailyfinance.com/2011/02/02/women-are-from-amazon-men-are-from-geico-worlds-most-desired/.

73. Clark, *A Farwell to Alms*.

74. Adrian Gallop, *Mortality Improvements and Improvement of Life Expectancies* (London: UK Government Actuary's Department, 2006).

75. Ridley, *The Rational Optimist*.

76. E. P. Thompson, "Time, Work-Discipline and Industrial Capitalism," *Past and Present* 38 (1967): 56–97.

7. HOW HUMANS ADAPT TO THE MODERN ENVIRONMENT

1. Clark, *A Farewell to Alms*.

2. D. M. Byrne, J. G. Fernald, and M. B. Reinsdorf, "Does the United States Have a Productivity Slowdown or a Measurement Problem?" draft, Brookings Papers on Economic Activity, BPEA Conference, March 10–11, 2016.

3. Gallop, *Mortality Improvements and Improvement of Life Expectancies*.

4. L. Cordain et al., "Physical Activity, Energy Expenditure and Fitness: An Evolutionary Perspective," *International Journal of Sports Medicine* 19, no. 5 (1988): 328–35.

5. A. Sclafani and D. Springer, "Dietary Obesity in Adult Rats: Similarities to Hypothalmic and Human Obesity Syndromes," *Physiology and Behavior* 17 (1975): 461–71.

6. U.S. Department of Health and Human Services, *2008 Physical Activity Guidelines for Americans* (Washington, DC: U.S. DSSH, 2008), https://health.gov/guidelines/guidelines/summary.aspx.

7. Wilkinson and Pickett, *The Spirit Level*.

8. D. Symons, "On the Use and Misuse of Darwinism in the Study of Human Behavior, Adaptiveness and Adaptation," *Ethology and Sociobiology* 17 (1992): 427–44.

9. Barber, "Why Behavior Matches Ecology."

10. Ibid.

11. Henrich, "Demography and Cultural Evolution."

12. K. M. Cummings, A. Brown, and R. O'Connor, "The Cigarette Controversy," *Cancer Epidemiology Biomarkers and Prevention* 16 (2007): 1070–76.

13. H. Barry, I. Child, and M. Bacon, "Relation of Child Training to Subsistence Economy," *American Anthropologist* 61 (1959): 51–63; J. W. Berry, "Independence and Conformity in Subsistence-Level Societies," *Journal of Personality and Social Psychology* 7 (1967): 415–18.

14. Barber, *Why Atheism Will Replace Religion*.

15. Chayanov, *The Theory of Peasant Economy*.

16. Low, "Cross-Cultural Patterns in the Training of Children."

17. Jiang and Sanchez-Barricarte, "Bride Price in China."

18. Barber, "Creative Productivity and Marriage Markets."

19. Caplow et al., *The First Measured Century*.

20. Hitsch et al., "What Makes You Click?"

21. Guttentag and Secord, *Too Many Women*.

22. L. Buckle, G. G. Gallup, and Z. A. Rodd "Marriage as a Reproductive Contract: Patterns of Marriage, Divorce, and Remarriage," *Ethology and Sociobiology* 17, no. 6 (1966): 363–77.

23. Wilson, *When Work Disappears*.

24. Inglehart and Welzel, *Modernization, Cultural Change*, 114.

25. Goldin, "Career and Family."

26. Barber, *The Science of Romance*.

27. M. Anderson, "Men Catch Up with Women on Overall Social Media Use," Pew Research Center, August 28, 2015, http://www.pewresearch.org/fact-tank/2015/08/28/men-catch-up-with-women-on-overall-social-media-use/.

28. P. R. Grant and B. R. Grant, "Hybridization, Sexual Imprinting, and Mate Choice," *American Naturalist* 149, (1977): 1–28.

29. Barber, *Why Atheism Will Replace Religion*.

30. Inglehart and Welzel, *Modernization, Cultural Change*, 272.

31. Pinker, *The Blank Slate*.

32. Odling-Smee et al., *Niche Construction*.

33. F. Gstulp, F. Barret, F. C. Tropf et al., "Does Natural Selection Favour Taller Stature among the Tallest People on Earth?" *Proceedings of the Royal Society, B. Biological Sciences* 282, no. 1806 (May 7, 2015), https://www.doi.org/10.1098/rspb.2015.0211.

34. J. Money, "The Syndrome of Abuse Dwarfism (Psychosocial Dwarfism or Reversible Hyposomatotropism)," *American Journal of Diseases of Childhood*,131 (1977): 508–13.

35. Gstulp et al., "Does Natural Selection Favour Taller Stature."

36. L. L. Cavalli-Sforza, P. Menozzi, and A. Piazza, *The History and Geography of Human Genes* (Princeton, NJ: Princeton University Press, 1996).

37. Franklin et al., "Epigenetic Transmission of the Impact of Early Stress."

38. Delaney-Black et al., "Violence Exposure, Trauma, and IQ."

39. Velasquez-Manoff, *An Epidemic of Absence*.

40. Morgan and Rose, "How Genetic Is School Myopia?"

41. Floud et al., *The Changing Body*, 23.

42. Case and Paxon, "Stature and Status."

43. Richerson and Boyd. *Not by Genes Alone*.

44. Floud et al., *The Changing Body*.

45. Ridley, *The Rational Optimist*.

46. Floud et al., *The Changing Body*.

47. Barber, "Creative Productivity and Marriage Markets."

48. Whiting and Whiting, *Children of Six Cultures*.

49. Nigel Barber, *Kindness in a Cruel World: The Evolution of Altruism* (Amherst, NY: Prometheus, 2004); Dreher, "Why Did the People of Roseto Live So Long?"

50. Blumberg, *Basic Instinct.*

51. Barber, *Why Parents Matter.*

52. Pinker, *The Better Angels of Our Nature.*

53. Cashdan, "Attracting Mates."

54. Caplow et al., *The First Measured Century.*

55. Galor and Moav, "Natural Selection."

56. Barber, "Creative Productivity and Marriage Markets."

57. Floud et al., *The Changing Body.*

58. David C. McClelland, *The Achieving Society* (New York: The Free Press, 1961).

59. Inglehart and Welzel, *Modernization, Cultural Change.*

60. James A. Tainter, *The Collapse of Complex Societies* (Cambridge, UK: Cambridge University Press, 1990).

61. Callaway, "Blocking 'Happiness' Chemical."

62. H. Pearson, "Africa's Locust Crisis Worsens," *Nature* (2004), doi:10.1038/news040816-13.

63. H. Dingle, *Migration: The Biology of Life on the Move*, 2nd ed. (Oxford, UK: Oxford University Press, 2014).

64. Groves, "The Advantages and Disadvantages."

65. Kotkin, *The Rise of Post-Familialism.*

66. Neal Ferguson, *The Square and the Tower: Networks and Power from the Freemasons to Facebook* (New York: Penguin, 2018).

67. Denyer, "China's Scary Lesson to the World."

68. "Fake News and Botnets: How Russia Weaponized the Web," *The Guardian*, December 2, 2017.

69. De Garine, "The Trouble with Meat."

70. T. J. Hatton and T. E. Bray, "Long-Run Trends in the Heights of European Men, 19th–20th Centuries," *Economics & Human Biology* 8 (2010): 405–13.

71. Gallop, *Mortality Improvements.*

72. M. R. Rosenzweig, "Aspects of the Search for Neural Mechanisms of Memory," *Annual Review of Psychology* 47 (1996): 1–33.

73. Johnson, *Everything Bad Is Good.*

74. Dickens and Flynn, "Heritability Estimates."

75. Hemelrijk and Hildenbrandt, "Schools of Fish and Flocks of Birds."

76. P. Chester, "Worldwide Trends in Honor Killings," *The Middle East Quarterly* 17, no. 2 (1996): 3–11.

77. Guttentag and Secord, *Too Many Women.*

78. N. Barber, "Cross-National Variation in Attitudes to Premarital Sex: Economic Development, Disease Risk, and Marriage Strength," *Cross-Cultural Research* 52, no. 3 (2018): 259–73, DOI: 10.1177/1069397117718143.

79. N. Barber, "Explaining Cross-National Differences in Polygyny Intensity: Resource-Defense, Sex Ratio, and Infectious Diseases," *Cross-Cultural Research* 42 (2008): 103–17.

80. J. E. Uecker and M. D. Rengnerus, "Bare Market: Campus Sex Ratios, Romantic Relationships, and Sexual Behavior," *Sociological Quarterly* 51 (2010): 408–35.

INDEX